CRYSTALLIZATION PROCESSES UNDER HYDROTHERMAL CONDITIONS

STUDIES IN SOVIET SCIENCE

1973

STUDIES IN SOVIET SCIENCE

CRYSTALLIZATION PROCESSES UNDER HYDROTHERMAL CONDITIONS

Edited by
A. N. Lobachev
Director, Laboratory of Hydrothermal Synthesis
Institute of Crystallography
Academy of Sciences of the USSR
Moscow, USSR

Translated from Russian by
G. D. Archard

CONSULTANTS BUREAU • NEW YORK-LONDON

Library of Congress Cataloging in Publication Data

Main entry under title:

Crystallization processes under hydrothermal conditions.

(Studies in Soviet science)
Translation of Issledovanie protsessov kristallizatsii v gidrotermal'nykh usloviiakh.
Includes bibliographical references.
1. Crystallization—Addresses, essays, lectures. 2. Crystals—Growth—Addresses, essays, lectures. I. Lobachev, A. N., ed. II. Akademiia nauk SSSR. Institut kristallografii. III. Series.

QD921.I8413   548'.5   73-79420
ISBN 978-1-4684-7525-8   ISBN 978-1-4684-7523-4 (eBook)
DOI 10.1007/978-1-4684-7523-4

A. N. Lobachev was born in 1910 and studied in the Chemical Faculty of Gor'kii State University. He has worked in the Institute of Crystallography of the Academy of Sciences of the USSR since 1941, and at present is the Assistant Director of the scientific section and Director of the Laboratory of Hydrothermal Synthesis. Lobachev is also a member of the Commission on Crystal Growth of the International Union of Crystallographers.

The original Russian text, published for the Institute of Crystallography of the Academy of Sciences of the USSR by Nauka Press in Moscow in 1970, has been corrected by the authors for the present edition. This translation is published under an agreement with Mezhdunarodnaya Kniga, the Soviet book export agency.

ISSLEDOVANIE PROTSESSOV KRISTALLIZATSII V GIDROTERMAL' NYKH USLOVIYAKH
A. N. Lobachev
Исследование процессов кристаллизации в гидротермальных условиях
А. Н. ЛОБАЧЕВ

© 1973 Consultants Bureau, New York
Softcover reprint of the hardcover 1st edition 1973
A Division of Plenum Publishing Corporation
227 West 17th Street, New York, N.Y. 10011

United Kingdom edition published by Consultants Bureau, London
A Division of Plenum Publishing Company, Ltd.
Davis House (4th Floor), 8 Scrubs Lane, Harlesden, London, NW10 6SE, England

Preface

This collection contains the results of a number of investigations which have been carried out in the Hydrothermal Synthesis Laboratory of the Institute of Crystallography, Academy of Sciences of the USSR; it constitutes a continuation of an earlier collection which appeared in 1968: Hydrothermal Synthesis of Crystals.

Problems associated with the synthesis of oxides, simple and complex sulfides, carbonates, silicates, and germanates are considered, and a great deal of factual material relating to the growth of single crystals of some of these compounds on a seed is presented.

Some of the articles pay special attention to the kinetic aspect of the growth of crystals; the conditions of growth are related to the morphological characteristics of the growing faces, and the relationship between the habit of the crystals and the composition and constitution of the solutions is considered.

A fair number of articles are concerned with the crystallization of new compounds, most of which have now been synthesized under hydrothermal conditions for the very first time; these include ternary chalcohalides of composition $A^V B^{VI} V^{VII}$, zirconates, lithium silicates, and germanates.

The collection also contains a description of the apparatus used for precision measurements at high temperatures and pressures.

We hope that this publication will present a better idea of the special characteristics of the hydrothermal method of synthesizing and growing crystals, and will prove useful to all those interested in this field of knowledge.

A. N. Lobachev

Contents

Some Problems of Hydrothermal Crystallization

L. M. Dem'yanets and A. N. Lobachev

The hydrothermal method of growing single crystals has become widely accepted in the last few decades. At the present time compounds of practically all mineral classes, from native elements to complex silicate phases, are synthesized under hydrothermal conditions.

In view of the complexity of the direct observation of crystallization at high temperatures and pressures, the theoretical basis of hydrothermal processes has by no means yet been fully established.* The majority of papers devoted to hydrothermal investigations are mainly concerned with a description of the experiments. By analyzing the very extensive experimental material which has been gathered together in connection with both natural and artificial mineral formation, we may elicit a number of generalizations regarding the hydrothermal synthesis of various compounds and the growth of the corresponding crystals.

At the present time hydrothermal chemistry is broadly divided into four fields of study: 1) the physicochemical characteristics of various systems at high temperatures and pressures; 2) the simulation of natural processes; 3) the synthesis of various compounds; 4) the growth of single crystals and the study of their growth kinetics.

In this article the authors have set themselves the task of examining certain laws relating to the growth of crystals and to the

---

*In this article we shall only be concerned with crystallization at moderate parameters, viz., temperatures of 100-600°C and pressures of under 3000 atm.

1

choice of the most effective solvents to be used for this purpose. Problems of crystal synthesis will also be treated in a number of cases.

Let us consider the composition of the systems which have been studied under hydrothermal conditions. The simplest of these is one of the A−H_2O type, where A is a compound slightly soluble at room temperatures (the SiO_2−H_2O system), or one of the B−H_2O type where B is a readily soluble salt, base, etc. (the NaCl−H_2O, NaOH−H_2O systems). In investigations of a mineralogical or petrographical character, we are usually concerned with systems of the A−A'−...−A^N−H_2O type (with various numbers of components), where A, A', ..., A^N are the oxides of the corresponding elements or minerals. In actually growing the crystals, we are interested in systems of the A−B−H_2O type, where A is the compound which it is desired to obtain in the form of a single crystal and B is the basic component of the mineralizing solvent. In the hydrothermal process water is a necessary component of the system, and indeed it is often the principal component as regards relative content; salts, acids, alkalis, etc. may serve as mineralizers.

In work carried out on the hydrothermal synthesis of crystals over the last thirty years [26−29] the majority of research workers have used pure water as solvent, being chiefly concerned with experimental geology and geochemistry involving the stability of minerals in aqueous media at high temperatures and pressures. By way of example we may mention the study of many silicate systems of the MgO−SiO_2−Al_2O_3−H_2O; CaO−SiO_2−H_2O; CaO−Al_2O_3−SiO_2−H_2O; type, systems such as GeO_2−H_2O; Ln_2O_3−H_2O, and many others.

In any system containing water, a certain concentration of the components occurs in the dissolved state, changing the pH of the medium, the composition of the original liquid phase, and the properties of the solution. Hence it is only justifiable to speak of carrying experiments out in "pure water" up to a certain extent, since we are in fact using aquous solutions of salts, acids, bases, etc., at very low concentrations.

A variety of compounds may normally be synthesized in such solutions. In order to grow large single crystals, however, it is customary to employ systems with a high concentration of a readily-soluble component.

TABLE 1

Class of compounds	Examples of compounds	Optimum solvent	Literature cited	Class of compounds	Examples of compounds	Optimum solvent	Literature cited
Native elements	Ag	MeOH	[1]	Phosphates	$AlPO_4$	H_3PO_4	[15]
	Se	MeOH	[2]	Silicates	Nepheline $Na[AlSiO_4]$	MeOH	[16]
	Pb	MeOH	[3]		Cancrinite $Na_8[Al_6Si_6O_{24}] \times (OH)_2$	MeOH	[17]
	Cd	MeOH	[4]		Sodalite $Na_8[Al_6Si_6O_{24}] \times (OH)_2$	MeOH	[18]
Oxides	ZnO	MeOH	[5]		Cd-Silicates	MeOH	[19]
	SiO_2	MeOH	[6]		Mn-Silicates	MeOH	[19]
	GeO_2	MeOH	[7]		Zn-Silicates	MeOH	[19]
	Al_2O_3	Me CO$_3$	[8,9]	Germanates	Zn_2GeO_4	MeOH	[7]
	Al_2O_3	MeOH	[8,9]		Na_4GeO_4	MeOH	[7]
	PbO	MeOH	[4]		K — Zn- Germanates	MeOH	[7]
	CuO	MeOH	[10]		Na — Zn- Germanates	MeOH	[7]
	Cu_2O	MeOH	[10]	Complex oxides	$Y_3Fe_5O_{12}$	MeOH	[20]
	Fe_3O_4	MeOH	[11]		$Me_3^{2-} Me_2^{3+} Me_3^{4+} O_{12}$	MeCl	[21]
	Fe_3O_4	MeCl	[12]	Sulfides	ZnS	MeOH	[22]
Carbonates	$CaCO_3$	MeCl	[13]		ZnS	Me_2S	[23]
Tungstates	$CdWO_4$	MeCl	[4]		FeS	MeCl	[24]
	$SrWO_4$	MeCl, MeOH	[4]		PbS	MeCl	[23]
	$BaWO_4$	MeCl, MeOH	[4]	Fluorides	CaF_2	MeCl	[25]
	$LiLn(WO_4)_2$	MeCl	[14]				
	$(Na, K) \times Ln(WO_4)_2$	MeCl	[14]				
Molybdates	$SrMoO_4$	MeCl, MeOH	[4]				
	$BaMoO_4$	MeCl, MeOH	[4]				
	$CdMoO_4$	MeCl	[4]				
	$PbMoO_4$	MeCl	[4]				

Published data [1-25] indicate that the composition of the effective solvents used in growing crystals is fairly standard. The most useful results are obtained on using one of a fairly restricted range of solvents, as indicated in Table 1.

The data presented in Table 1 by no means represent every type of solvent used by the original authors in synthesizing and recrystallizing various compounds. In the table we have simply included those solvents which lead to the recrystallization of the original material in such a way as to yield a single-mineral product, and in the majority of cases growth on a seed crystal.

On the basis of general physicochemical considerations we

may define a number of particular conditions which have to be satisfied in order to select an effective solvent for growing crystals:

1. Congruence of the dissolution of the test compounds.
2. A fairly sharp change in the solubility of the compounds with changing temperature or pressure,
3. A specific quantitative value of the absolute solubility of the compound being crystallized;
4. The formation of readily-soluble mobile complexes in the solution,
5. A specific redox potential of the medium, ensuring the existence of ions of the required valence.

One of the most important questions is the effect of the mineralizer* B on crystallization in the system $A-B-H_2O$, this effect being associated with the action of one particular type of ion, with changes in the ionic strength, the structure and properties of the solution, and so on. This is a problem requiring very special consideration, and we shall not be concerned with it in this article. We have also deliberately refrained from considering various other questions specifically related to the character of the solvent. These include the structure of aqueous solutions, the change in the structure of the solution with increasing concentration of the solvent, the size of the particles in various solutions, the phenomena of polymerization and depolymerization, the existence of active centers in solutions, and so on. We shall now consider the principal requirements imposed upon solutions to be used in the synthesis and growth of crystals.

1. Congruence of Dissolution of the Test Compound

Various compounds may be synthesized even under conditions of incongruent dissolution as a result of the partial or complete decomposition of the principal crystalline phase and the formation of a number of others. By way of example we may mention the systems $CaWO_4- meOH-H_2O$, $CdMoO_4-LiCl-H_2O$, $Na_8[Al_6Si_6O_{24}]$ $(OH, Cl)-NaOH-H_2O$: 1) Owing to the decomposition of calcium

---

*In future we shall regard "mineralizer" as meaning a substance introduced into the $A-H_2O$ system and not forming any poorly-soluble compounds with the components of A.

tungstate and an exchange reaction with alkali, $Ca(OH)_2$ is formed
[30]; 2) owing to the decomposition of $CdMoO_4$ by concentrated
aqueous solutions of LiCl, and Li—Cd molybdate is formed [4]; 3)
in aqueous solutions of NaOH (10-15% concentration), hydrocancrinite and other aluminosilicates are formed during the decomposition of sodalite.

Separate consideration is really needed for the synthesis of
poorly-soluble compounds from composite components (for example,
the synthesis of beryl and tourmaline in B-containing media) when
crystals may be grown on a seed by virtue of the dissolution of the
oxides (or corresponding minerals), and growth ceases at the end
of the synthesis [31, 32]. Recrystallization of beryl and tourmaline
does not take place under these conditions. Crystals may clearly be
grown from these compounds as a result of synthesis provided that
there is a constant and uninterrupted supply of material to the growth
zone, such as may be obtained in a flow-type installation.

An interesting example of crystallization under conditions of
incoherent dissolution is the crystallization of yttrium ferrogarnet.
According to [12], synthesis of the garnet takes place in aqueous
solutions of Na_2CO_3, the character of the process changing with
time (Fig. 1). The maximum yield of garnet is secured if the conditions are maintained for seven or eight days; over a longer period
the garnet suffers partial decomposition, with the formation of mag-
netite and the orthoferrite $YFeO_3$. Of course it is quite possible
that in this case the change in the phase composition of the system
with time is also associated with purely technical causes; the inad-
equate sealing of the vessels, or the partial reduction of the Fe^{3+}
by hydrogen evolving during the corrosion of the autoclave walls
(the presence of Fe^{2+} may change the course of the reaction in which
the garnet is formed).

In a number of cases material crystallizes on a seed as a
result of simultaneous synthesis and recrystallization processes,
under conditions promoting the congruent dissolution of the crys-
taliizing compound. By way of example, we may mention the
synthesis and recrystallization of the D phase Na_2ZnGeO_4 in the
Na_2O—ZnO—GeO_2—H_2O system [7], and the crystallization of an
Li—Cd molybdate of variable composition $LiCd(MoO_4)(OH)$ in the
CdO—MoO_3—$LiCl$—H_2O system when there is a large excess of the
first component.

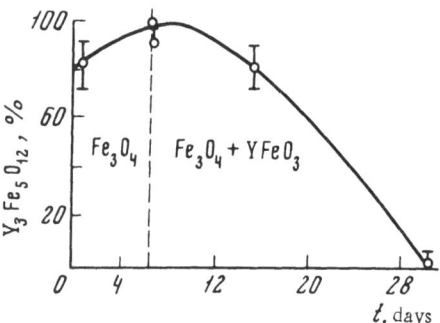

Fig. 1. Time dependence of the yield of garnet
in the system $Y_2O_3-Fe_2O_3-Na_2CO_3-H_2O$ [12].

In order to grow single crystals of various materials it is
usual to employ recrystallization of the original material. In this
case, in contrast to that of synthesis, it is essential that the compound under consideration should dissolve congruently in the solvent chosen.

2. Change in the Solubility of the Compound with Temperature and Pressure

It is well known that the majority of compounds may be divided
into two classes as regards the nature of their solubility in water.

The first type include compounds in which the solubility increases continuously with rising temperature up to the melting
point. For aqueous solutions of this kind of compound there are
no critical phenomena in the state of saturation. By way of
example we may mention solutions of alkali halides (except NaF
and LiF) in water, a number of bromides, iodides, nitrates, and
thiocyanates of alkali metals and ammonia, and so on [33-36].
In saturated aqueous solutions of these salts there is a continuous
change in the equilibrium vapor pressure with rising temperature.
For such systems the three-phase equilibrium (Fig. 2) curve is
characterized by a maximum, the temperature T_{max} of this
maximum being related to the melting point T_m by the empirical
equation [34]

$$\frac{1}{T_{max}} = \frac{1}{T_m} + 0.00021.$$

The higher the melting point of the salt (other conditions being equal), the higher are the temperature and pressure corresponding to the maximum on the curve of three-phase equilibrium. For more soluble salts, smaller maximum pressures occur.

In systems of the first type, critical phenomena only appear in unsaturated solutions; hence in the temperature/composition and pressure/composition cross sections the critical curve does not intersect the curve of compositions of the saturated solutions, and there is a continuous transition from the solutions of the melt (Fig. 2).

Systems of the second type include compounds with a low solubility in water near the critical temperature of the latter. These are compounds readily soluble in water at room temperature but having a negative solutibility coefficient (Na_2CO_3, Na_2SO_4 etc.) [37, 38], and also compounds only slightly soluble in water at both room and higher temperatures (oxides and salts of refractory metals, sulfides, silicates, and salts of a number of oxygen-containing acids).

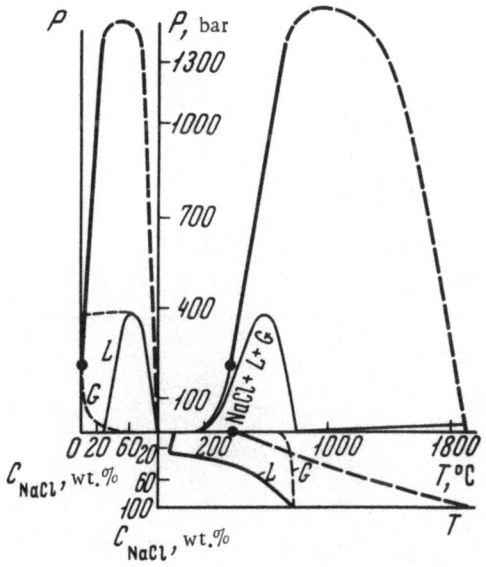

Fig. 2. Example of a system of the first type, the NaCl–H_2O system. The critical points are shown on the curves.

Systems of this type are characterized by the fact that, in
view of the low absolute value of the solubility, critical phenomena
occur even in saturated solutions. We see from Fig. 3 that the
critical curve and the curve of three-phase equilibrium intersect
at two points p and Q constituting critical points of the saturated
solutions. Between the first and second critical points is a region
of fluid solutions (region of divariant equilibrium). Above the
second critical point these systems are analogous to systems having no critical phenomena in saturated solutions.

There is as yet no exact theory explaining the influence of
temperature on the solubility of compounds in water. The temperature dependence of solubility was considered in general form by
Chin [39, 40]. A rise in temperature leads to an increase in the
water-solubility of a relatively small group of salts (chlorides and
other compounds readily soluble under ordinary conditions). For
the majority of refractory compounds the water-solubility falls
with increasing temperature (in pure water). For salts with ions
which undergo strong hydration, a negative coefficient of solubility
is more likely [39, 40].

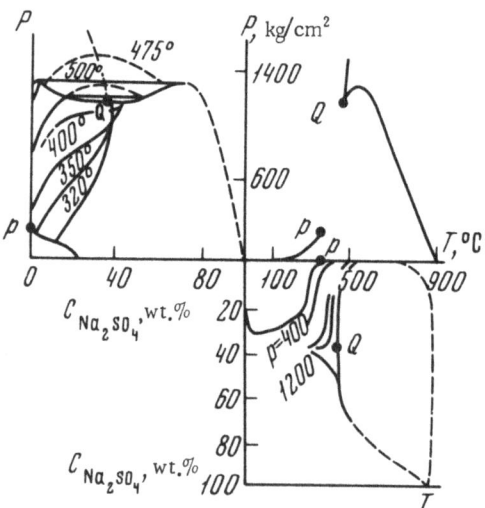

Fig. 3. Example of a system of the second type
with two critical points P and Q, the Na_2SO_4–H_2O
system [37].

It should be noted that the sign of the change in solubility is by no means always the same over a wide range of temperatures and pressures. The compound under consideration may be characterized by a change in the sign of the temperature coefficient of solubility, depending on the concentration of the solution. The manner in which the solubility varies may be completely different in relatively weak and concentrated solutions. In a number of cases, for small concentrations of the mineralizer, the compound has a negative temperature coefficient of solubility, and for higher concentrations a positive one. This kind of solubility was found [41-44] in the systems K_2SO_4–KCl, K_2SO_4–$NaCl$, K_2SO_4–$NaOH$, Na_3PO_4–$NaOH$, Na_2SO_4–KCl, Na_2SO_4–$NaCl$, Na_2SO_4–$NaOH$, Na_3PO_4–$NaCl$ etc. This problem is related to the structure and properties of solutions of different concentrations, and we shall not consider it in any great detail.

The use of compounds with the first type of solubility as hydrothermal solvents enables us to study crystallization (or solubility) over a wider temperature and concentration range, since the introduction of an easily soluble compound increases the critical temperature of the solution by comparison with that of water.

The influence of pressure on the solubility was first considered by Adams and Gibson [45, 46]. It was pointed out in these papers that the pressure effect (the fall or rise in solubility with increasing pressure) was associated with the sign of Δ ($\Delta = \overline{V} - V^s$), where $\overline{V}$ is the partial molar volume of the salt in the saturated solution, V^s is the molar volume of the solid salt at the same temperatures and pressures. With increasing pressure the partial molar volume of the solid salt diminishes, and the sign may even reverse. Equally between $\overline{V}$ and V^s will correspond to the maximum on the solubility curve. If the energy of interaction between the salt particles and the water is high at low temperatures and the solubility of the salt is low, an increase in pressure usually leads to a rise in solubility [47].

3. Absolute Solubility of the Crystallizing Compound

For the growth of single crystals under hydrothermal conditions to take place continuously in time, it is essential that there should be a constant flow of supply material into the growth zone, in quantities sufficient to provide for a specific rate of growth on the seed v = kΔS where v is the rate of growth of the crystal and ΔS is the supersaturation in the growth zone.

The growth rates of single crystals under conditions of fixed supersaturation have been measured for a comparatively small number of compounds. Let us briefly consider the most characteristic examples.

It is important to emphasize that we are considering the relation between the growth rates and the supersaturation in the growth zone for the case of crystals growing on a seed; for spontaneous crystallization the growth rate may be much higher.

According to the results of [48], the solubility of SiO_2 in 1 and 13% NaOH solutions respectively equals 0.15 and 4 wt.% at 250°C and 0.3 and 4.8% at 450°C. For this value of the solubility a fair degree of supersaturation is achieved in the growth zone, and hence this method of growth is of specific practical interest.

According to the results of [49], the solubility of quartz in 0.5 M NaOH (for an occupation factor of f = 0.8) varies between 2.13 and 2.4 g per 100 g of solvent over the temperature range 300-400°C. Under these conditions the maximum growth rate occurs on the (0001) face. For a temperature of 385°C and f = 0.81 we have the following growth rates on the faces: (0001) ~ 2.4 mm/day, (01$\bar{1}$1)–1.1 mm/day, (10$\bar{1}$1) ~ 0.25 mm/day. With increasing occupation factor the growth rates of the rapidly-growing faces reach approximately 5 mm/day. Such values are too high for the growth rates of crystals from solutions and approach those encountered in melt systems. For hydrothermal systems rates of this order are not very common.

As noted in [50], a solubility of zincite comparable with that of quartz is obtained for NaOH concentrations of the order of 4-6 M. The solubility of ZnO in 6.24 M NaOH solution is approximately 4.5-6 wt.% between 200 and 400°C. The growth rate of zincite crystals in a direction perpendicular to the (0001) faces at ΔT = 45° is 0.25 mm/day.

According to [9, 51], the growth rate of corundum crystals on the (11$\bar{2}$0) faces in a 10-15% solution of Na_2CO_3 is approximately 0.25-0.3 mm/day; the growth rate of the ($10\bar{1}1$) faces in $NaHCO_3$ solutions reaches 1 mm/day.

It was noted in [18] that the growth rate of sodalite on a seed under optimum growth conditions was approximately 0.3-0.4 mm/day for $\Delta T = 15°$, a temperature of 270°C, and an NaOH concentration of 30 wt.%.

According to [7], the growth rate of Na_2ZnGeO_4 crystals under optimum conditions equalled approximately 0.4-0.6 mm/day in a 30% aqueous solution of NaOH at $\Delta T = 15°$.

Some data relating to the growth rates of individual compounds are presented in Table 2.

We see from Table 2 that in all the cases indicated the solubility of the compounds reaches a value of the order of several wt.%.

The supersaturation in the growth zone when growing high-quality crystals is as follows: for quartz 0.07 wt.% in 0.51 M NaOH, for zincite, 0.1 wt.% in 6.47 M KOH, for sodalite, $\sim$ 0.2 wt.% in 30% NaOH [6, 52-54].

The supersaturation in the growth zone needed for growing good quality crystals depends on the properties of the solvent and also those of the substance under consideration; it is related to the development of crystallization centers in the growth zone, surface phenomena on the seed, the convection velocity, and other factors.

TABLE 2

Compounds	Solvent	Solubility, wt.%	T_{opt}, °C	P_{opt}, atm	Literature cited
Zincite ZnO	6.47 M KOH	4.5	350	588	[52]
	6.24 M NaOH	5.7	350	588	[52]
Quartz SiO_2	0.51 M NaOH	3.0	350	1480	[6]
Sodalite $Na_8[Al_6Si_6O_{24}](OH)_2$	6.4 M NaOH	2.2*	300	400	[53]
	10.8 M NaOH	3.2	300	400	[53]
Corundum Al_2O_3	0.5 M NaOH	2.0*	400	680	[54]

*Not optimum growth conditions; these data are presented for comparison with the solubility of the other compounds.

Thus for different substances the values of ΔS will differ (in the same way as the temperature, pressure, and composition of the solvent); our results cannot therefore claim to represent all possible hydrothermal systems.

The foregoing brief review of some of the systems in question leads to the conclusion that, for the practical growth of single crystals under hydrothermal conditions, it is in general essential that the material in question should dissolve in the selected solvent to the extent of a few weight percent. This order of solubility enables us to create a supersaturation in the growth zone sufficient to produce a high-quality crystal at an acceptable velocity (~ 0.2-1 mm/day) and at a moderate temperature.

4. Complexing in Hydrothermal Solutions

The principal materials employed in the hydrothermal synthesis of crystals are compounds of the second class of water-solubility. These are compounds which are practically insoluble in water at room temperature, and which are difficult (or impossible) to obtain in the form of single crystals by any other growth methods.

The method usually employed to obtain compounds under hydrothermal conditions is the temperature drop method, in which the growth of the substance arises mainly from a steady flow of supply material, conveyed to the growth zone by convection currents. Crystal growth is based on the dissolution—transfer—crystallization principle, in which the middle stage, the transfer of the material, plays an essential part. This is essentially a case of transport reactions in which the mineralizer introduced into the system is the transporting agent. Chemical reactions are usually involved in the mass transfer. One of the necessary conditions of transport is the existence of a concentration gradient created by a temperature gradient along the autoclave.

Repeated experiments and thermodynamic calculations have shown that pure water cannot serve as an essential transporting agent under hydrothermal conditions [55], since most refractory compounds are almost insoluble in water, and their solubility rises very little with temperature (or in certain cases even diminishes). In order to increase the solubility it is essential to introduce a mineralizing substance forming mobile complexes to play the part of transporting agent.

The formation and stability of the complexes depend on the composition of the system, the temperature, pressure, and a number of other factors. Under equilibrium conditions, chemical equilibrium may be supposed to exist between the complexes and the free ions in aqueous electrolyte solutions.

It is now generally accepted that metal ions form aqua-complexes in solutions [56]. The number of water molecules in the inner sphere of an aqua-complex usually equals the maximum coordination number of the cation. In the outer sphere of the complex, water dipoles occur; the number of spheres around the central atom is three or over. The complex may be mononuclear or polynuclear. In the majority of inorganic systems containing water, polynuclear complexes are rare [57, 58].

In continuing our comparison between hydrothermal and transport reactions, we may consider the transport of a number of elements in the latter in the presence of traces of water.

One example of such chemical transport reactions is the transport of tungsten in the following manner [59, 60]:

$$W + x H_2 O_{gas} = WO_{x gas} + x H_2 \ (2400° \ C),$$
$$WO_3 + H_2 O_{gas} = WO_2 (OH)_2 \ (1100° \ C),$$

another is the transport of beryllium [61]:

$$BeO + H_2 O = Be (OH)_{2 gas} \ (1500° \ C).$$

A classical example of transport in the presence of water is the transport of SiO_2 in chemical transport reactions and hydrothermal conditions [62]:

$$SiO_{2 sol} + n H_2 O_{gas} = SiO_2 \cdot n H_2 O_{gas}.$$

This example clearly supports the analogy between chemical transport reactions and the transfer of elements under hydrothermal conditions.

The values obtained for the solubility of quartz in water suggested that the dissolved SiO_2 molecules combined with H_2O to form $Si(OH)_4$ [63]. Some authors have acknowledged the possibility that $SiO(OH)_2$ and other complexes may also be formed. The important fact for our own case is that, firstly, in the presence of water or water vapor silicon may be transported in the form of specific complex compounds, and, secondly, silicon combines with water to form aqua-complexes of the $SiO_2 \cdot nH_2O$, $SiO(OH)_2 \cdot nH_2O$ and other types.

By analogy with the reactions for quartz and sapphire, the following reaction was given in [50] for the dissolution of zincite in water:

$$ZnO_{sol} + nH_2O \rightleftharpoons ZnO \cdot nH_2O.$$

The same results are obtained for the dissolution of other oxides (for example, Al_2O_3 and GeO_2) in water (or aqueous solutions).

If a solution contains additional complexing ions, there may be a regrouping of the water complex, with the partial or complete loss of water in the first coordination sphere and the formation of amore complicated complex.

The transport of various substances in the form of complex compounds with such anions as F^-, Cl^-, CO_3^{2-}, and others has now been firmly established for chemical transport reactions, and in a number of cases for hydrothermal processes as well. We shall shortly consider some examples of such reactions.

Anions play the chief part in the transport of the chemical elements, but a certain contribution from the cations must also be taken into account.

Thus it is pointed out in [64, 65] that the influence of the outer-sphere cations on complexing lies in a change in the number of ligands attached to the central ion, or a change of strength in the metal-ligand bond. It is shown in [66] that the influence of such cations on the state of complexing of the central ion is associated with the different strengths of the ion pairs formed between the complex anion and the outer-sphere cation.

The effect of alkali metals (from Li to Cs) on the state of complexing of cadmium (at room temperature) was considered in

[67]. The tendency toward association with the complex anion was smallest in the case of the lithium ion; in the presence of lithium the following reaction occurred:

$$Cd^{2+} + nCl^- \rightleftharpoons CdCl_n^{2-n}.$$

If some of the lithium in the system is replaced by another alkali element B, then the following reaction also occurs:

$$mB^+ + CdCl_n^{2-n} \rightleftharpoons B_mCdCl_n.$$

The lithium (less strongly bound to the complex) is replaced by the other alkali metal, and the degree of complexing therefore increases. Complexes of the following types inter alia are thus formed: $K(H_2O)CdCl_4$, $K_2(H_2O)_x CdCl_4$. As temperature increases, this relationship is not always preserved; for example, in chloride solutions mixed lithium-cadmium compounds of ultimate composition $LiCdMoO_4OH$ are formed. In individual cases the relationships in question may nevertheless remain intact (for example, in the crystallization of mixed tungstates of monovalent metals and rare-earth elements).

Complexing under hydrothermal conditions was considered in greatest detail in [55].

A large number of natural vein and pegmatitic minerals were studied in [68], and a scheme was proposed for distinguishing the elements according to their behavior in alkaline and acid solutions (Table 3).

In our own opinion this scheme is only applicable in the absence of complexing ions; in the presence of such ions, however, there should be a state of equilibrium between the developing complexes and the corresponding simple ions.

A number of authors consider that, in individual cases, the transport of ore elements under hydrothermal conditions may be explained without having recourse to the concepts of easily-soluble, simple or complex compounds. The transport of the elements may, according to these authors, be explained by the high solubility of the substances in question in comparatively concentrated aqueous solutions of electrolytes [9].

TABLE 3

Li, Na, K, Ba, Sr, Ca, Mg	Fe²⁺, Mn, Al, Zr, Ti, Be, Ce, Fe⁺³, Sn, Ta, Nb, W		P, Si, C, B, S, Cl, F	
Basic oxides	Amphoteric oxides		Acid oxides	
	more basic	more acid		
Transported as simple ions in alkaline solutions	Transported as complex oxygen ions in alkaline solutions		Transported as simple ions	
Transported as simple ions in acid solutions		Transported as complex ions in acid solutions	Transported as simple ions in acid solutions	

5. Redox Potential of the System

Many elements are characterized by a variable valence, and redox reactions taking place in aqueous solutions therefore assume an important role. Unfortunately, existing values of redox potentials have been mainly determined at room temperature and atmospheric pressure; if these are used for calculating reactions under hydrothermal conditions without introducing proper corrections, there may in a number of cases be a considerable error, the extent of this being difficult to estimate.

The oxidation potential at a temperature differing from 25°C may be determined from the expression

$$E = E_0 + \frac{RT}{n_e F} \ln Q,$$

where E_0 is the standard electrode potential at 25°C, R is the gas constant, T is the absolute temperature, F is the isochore-isothermal potential or free energy, n_e is the number of electrons taking part in the reaction, Q is the fraction obtained on dividing the product of the activities of the reaction products by the product of the activities of the reagents (the value of each activity being raised to a power equal to the coefficient of the compound in the reaction in question). If the solution is reasonably dilute, the concentrations of the components may be used in the equation instead of the activities.

A substantial temperature rise primarily affects the activity ratio of the components, and this may lead to a considerable change in the redox potential.

The quantity E is closely related to the change in the pH of the solution [70, 71]. Usually E falls with increasing pH. For a number of reactions the E = f(pH) curves contain regions in which E is almost independent of pH (Fig. 4). For example, in strongly acid solutions ferrous oxide occurs (E = 0.77 V). At pH = 2-3 the value of E falls sharply and becomes negative, and in solutions weakly-acid enough for the precipitation of $Fe(OH)_3$ an oxidation process $Fe^{2+} \rightarrow Fe^{3+}$, takes place; hence compounds of ferrous oxide may be obtained in strongly acid media [70].

The electrode potentials for reactions taking place in aqueous solution should be limited by the values of E relating to the decomposition of water:

$$2H_2O = O_2 + 4H^+ + 4\bar{e}, \; E_0 = 1.228 \; V,$$
$$H_2 = 2H^+ + 2\bar{e}, \; E_0 = 0 \; V.$$

If the potential of any reaction exceeds 1.228 V, theoretically the water should decompose with the evolution of oxygen; if E < 0

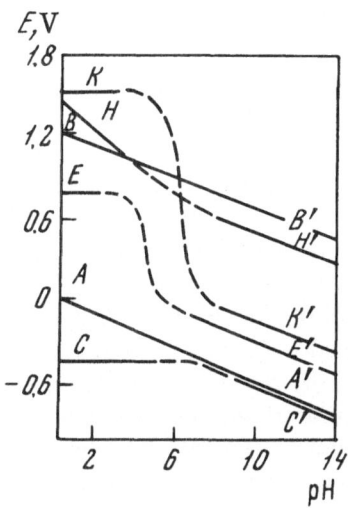

Fig. 4. Dependence of the redox potential of certain reactions on pH [70] A—H_2 = $2H^+ + 2e$; A'—$H_2 + 2OH^- = 2H_2O + 2e$; B—$2H_2O = O_2 + 4H^+ + 4e$; B'—$4OH^- = O_2 + 2H_2O + 4e$; C—$Fe = Fe^{+2} + 2e$; C'—$Fe + 2OH^- = Fe(OH)_2 + 2e$; H—$Pb^{+2} + 2H_2O = PbO_2 + 4H^+ + 2e$; H'—$PbO + 2OH^- = PbO_2 + H_2O + 2e$; K—$Mn^{2+} = Mn^{3+} + e$; K'—$Mn(OH)_2 + OH^- = Mn(OH)_3 + e$; E—$Fe^{2+} = Fe^{3+} + e$; E'—$Fe(OH)_2 + OH^- = Fe(OH)_3 + e$.

the water should decompose with the evolution of hydrogen. In practice higher potentials are required for the decomposition of water in view of the overvoltage effect; nevertheless, the potentials of the two reactions indicated largely determine the redox processes in solutions.

The relation between the redox potential of the medium and natural mineral formation or the formation of artificial minerals was considered in [70, 72, 8]. It was found in the cases of corundum and ruby that an increase in the redox potential of the medium was accompanied by an increase of the amount of chromium in solution and a corresponding reduction in the amount of chromium in the crystals. The redox potential is one of the most important factors determining the capture coefficient of chromium by the corundum crystals; this is because of the change in the valence state of chronium in solution which takes place with changing pH and Eh. A reduction in the Eh of the medium leads to a successive substitution of the complexes CrO_4^{2-}, $Cr_2O_7^{2-}$, CrO_2^-, Cr^{+3}, Cr^{+2}. The optimum conditions for the growth of ruby correspond to the range of Eh values between -0.12 and +0.01 V, for which CrO_2^- (analogous to the aluminate complex AlO_2^-) is probably a stable complex [8].

A change in the pH of the medium affects the degree of formation of various complexes, and also their composition. By way of example, we may mention the composition of Mo and W complexes in aqueous solutions.

In strongly acid media, Mo^{6+} and W^{6+} form cations, and in alkaline media anions of monomolybdic and monotungstic acids [73, 74]. The composition of the complex Mo and W anions takes a large number of forms. For every specified pH value there is an equilibrium system with a predominant number of ions of one type or another. In a strongly acid medium the Mo^{6+} cations are either weakly polymerized or remain monomeric. In this case the state of the Mo^{6+} depends on the form of the acid: In the presence of HCL molybdenyl cations are formed over a narrow range of HCl concentrations (2-0.2 N); on increasing the concentrations, complex chloride ions occur. In sulfuric acid complex anions predominate in every case, while in HNO_3 and $HClO_4$ cations tend to dominate [75]. According to [76] polymeric forms develop at pH = 6.5, the most stable of these being the hexamolybdate ion

$[Mo_6O_{20}]^{4-}$. The same ion exists on acidifying a solution of alkaline tungstate. On further acidification the ion transforms into the paratungstate $[H_3W_6O_{21}]^{3-}$, and so on. In alkaline solutions the ions $[MoO_4]^{2-}$ and $[WO_4]^{2-}$ exist.

The change in the composition of germanate and silicate ions with pH value is considered in [7, 18].

The pH of the medium displaces the boundary of phase transformations and changes their velocity [77]. For example, an increase in pH leads to an increase in the rate at which $Al_2O_3 \cdot 1.3H_2O$ transforms into diaspore and corundum at 350 and 400°C. A change in pH affects the hydrothermal transformation of quartz glass into cristobalite and quartz, boehmite into diaspore, and so on.

Repeated experiments and thermodynamic calculations have shown that natural hydrothermal ore-forming solutions are weakly-dissociated alkaline chloride solutions of electrolytes. At low and hypercritical temperatures the pH value is almost neutral, and varies approximately within the range 3-9. At medium temperatures of ore formation the solutions are probably more acid [55].

In the artificial growth of crystals the range of variation of the pH value is much wider. Crystals of various minerals may thus be grown both in very acid and in neutral or alkaline solutions, depending on the individual characteristics of the particular mineral.

*　　*　　*

In hydrothermal synthesis the solvents employed are usually compounds of the first type of solubility, their components forming a variety of complexes with the substance under consideration. We may logically divide the solvents usually employed in the hydrothermal process into the following groups:

    (1)　halides of the alkali metals and ammonia, as well as of a number of divalent metals;

    (2)　hydroxides of the alkali metals;

    (3)　salts of weak acids (H_2CO_3, H_3BO_3, H_3PO_4, H_2S) and alkali metals;

    (4)　salts of strong acids;

    (5)　acids.

This classification is based principally on the capacity of
the anions (or cations) in the mineralizers to form complex compounds by changing the activity of the components of the dissolving
compound, and also on the oxidizing characteristics of the medium.

Practice shows that the most effective solvents are compounds
of the first two groups (halides and hydroxides of the alkali metals)
(Table 1), and these we shall consider in more detail.

In the first group, fluorine, as a universal addend, forms
strong complexes with many elements, and these are fairly stable
even under hydrothermal conditions. The formation of such complexes may greatly change the activity of the central atom (silicon
in $Na_2[SiF_6]$, beryllium in $Na_2[BeF_4]$, aluminum in $Na_3[AlF_6]$, etc.).
Within the range of stability of the complex ions at high temperatures and pressures, the activity of the central atom may be close
to zero, and the element in question will then be taken out of the
reaction, which is not always desirable. An example of this is the
partial binding of aluminum into fluoride complexes in the synthesis of tourmaline, when the aluminum precipitates in the form
of fluoroaluminates.

Complexes with HCl, NaCl, and KCl are mainly formed at
high temperatures (above 200°C). The stability of many chloride
complexes of metals increases with rising temperature. Various
Cl complexes have been studied in greatest detail [55, 56, 78].

With increasing temperature the polar properties of liquid
water are so weakened ($\varepsilon \simeq 10-11$) that the dissociation of such
electrolytes as HCl and NaCl may be strongly suppressed, particularly for high concentrations of the solution. For example,
at 450°C potassium chloride behaves as a weak electrolyte. Hence
even the strongest electrolytes will occur in solution chiefly as ion
pairs or larger formations [79].

At high temperatures HCl is a relatively stable complex.
An increase in the chloride concentration in the solution increases
the extent to which this complex is formed; on reducing the temperature the HCl dissociates. In aqueous solutions of chlorides,
in the presence of, for example, silver, the following types of complex are formed: $AgCl_2^-$, $AgCl_3^{2-}$, $AgCl_4^{3-}$. Chlorine complexes
also occur for mercury (e.g., $HgCl_4^{2-}$, $HgCl^+$) [55]. For lead the

most important complex is $PbCl^+$; at temperatures below 200°, $PbCl_4^{2-}$ predominates in concentrated solutions. Chlorine complexes are found in the case of cadmium (in the synthesis of CdS, chloride complexes of cadmium prevent the precipitation of the hydrothermal cadmium sulfide CdS), tungsten, molybdenum, tin, and others. It is interesting to note the analogy between the forms of transport in hydrothermal and chemical transport reactions. For example in transport reactions [60] tungsten (and molybdenum) is transported in the form of WCl_6. Analogous complexes were also found in aqueous solutions of chlorides, which indicates the wide temperature range of existence of complexes of this kind.

By analogy with chemical transport reactions, we may consider that Fe, Zr, Nb, and Ni and other elements may be transported under hydrothermal conditions in the form of compounds with chlorine, bromine, and iodine.

The high absolute solubility of many compounds in aqueous solutions of alkali chlorides and the formation of mobile complexes under hydrothermal conditions are factors which may serve to explain the effective use of chloride solutions in obtaining single crystals of a number of compounds.

As already indicated, aqueous solutions of sodium, potassium, and lithium hydroxides are also widely used as optimum hydrothermal solvents.

Single crystals of native elements, oxides, and a number of salts of oxygen-containing acids may be obtained in aqueous solutions of alkalis. Solutions of alkalis are more effective for obtaining single crystals of various silicates, aluminosilicates, and germanates. At high temperatures bases become weakly dissociated, and may serve as specific centers of crystallization for compounds which have the developing group of atoms "inscribed" in their structure. This is why aqueous solutions of alkalis have proved so useful in producing sodium and potassium silicates and germanates.

As in pure water, aqua-complexes of the $SiO_2 \cdot nH_2O$; $SiO_a^{(2a-4)}$ type are formed in alkali solutions. The following scheme is proposed in [6] for the dissolution of quartz in hydrothermal alkali solutions:

$$2SiO_2 + H_2O + (2a - 4)(OH)^- \rightleftharpoons SiO_2 \cdot nH_2O + SiO_a^{(2a-4)} + (a - 2)H_2O.$$

Experiments show that for NaOH solutions $a = \frac{7}{3}$ and for Na_2CO_3 $a = 3$.

A similar reaction describes the dissolution of ZnO in alkali solutions [50]:

$$ZnO + (2a - 2)(OH)^- \rightleftharpoons ZnO_a^{(2a-2)} + (a - 1)H_2O.$$

Basing our considerations on the fact that, on passing into alkali solution, elements or groups of elements form aqua-complexes, we may write

$$SiO_2 + H_2O \rightarrow SiO_2 \cdot aq,$$
$$ZnO + H_2O \rightarrow ZnO \cdot aq,$$
$$GeO_2 + H_2O \rightarrow GeO_2 \cdot aq.$$

Considering that the coordination number of a particular element in solution is similar to its coordination number in the crystalline state, we may write

$$ZnO + H_2O \rightarrow [Zn(H_2O)_4]^{2+},$$
$$SiO_2 + H_2O \rightarrow [Si(H_2O)_4]^{4+},$$
$$GeO_2 + H_2O \rightarrow [Ge(H_2O)_4]^{4+},$$
$$Al_2O_3 + H_2O \rightarrow [Al(H_2O)_4]^{3+}.$$

In concentrated alkali solutions we find that, at room temperatures, at which the alkalis are strongly associated, complexes of the following types exist in these solutions: $Zn(OH)_4^{2-}$, $Si(OH)_4$, $Si(OH)_6^{2-}$, $Al(OH)_4^-$. The existence of such ions (for example, $Al(OH)_4^-$ in aluminate solutions) is supported by Raman spectroscopy and other experimental methods. On raising the temperature the following complexes are the most probable: $[Zn(H_2O)_x(OH)_{4-x}]^{x-2}$, $[Ge(H_2O)_x(OH)_{4-x}]^{+x}$; $[Si(H_2O)_x(OH)_{4-x}]^{+x}$; $[Ge(H_2O)_x(OH)_{6-x}]^{x-2}$; $[Al(H_2O)_x(OH)_{6-x}]^{x-3}$ etc. Such complexes have been found for aluminum [77]:

$$Al_2O_3 + 7H_2O + 2OH^- = 2Al(H_2O)_2 \cdot (OH)_4^-.$$

On adding alkali to the oxide of the element, the corresponding germanates, aluminates, zincates, etc. are formed, these existing in the form of intermediate or final reaction products. By way of example [7], we may mention crystallization in the

$Na_2O-ZnO-GeO_2-H_2O$ system:

$$GeO_2 + 2NaOH \rightleftarrows Na_2GeO_3 + H_2O,$$
$$ZnO + 2NaOH \rightleftarrows Na_2ZnO_2 + H_2O,$$
$$ZnO + 2NaOH + H_2O \rightleftarrows Na_2[Zn(OH)_4].$$
$$Na_2GeO_3 + 2Na_2ZnO_2 + 3H_2O \rightleftarrows ZnGeO_4 + 6OH^-,$$
$$Na_2GeO_3 + Na_2[Zn(OH)_4] \rightleftarrows Na_2ZnGeO_4 + 2OH^- + H_2O.$$

Questions as to the forms of existence and transport of various ions in hydrothermal aqueous solutions are complicated and controversial; factual information is as yet insufficient to ensure a unique solution.

Systems incorporating MeOH provide a typical example in which the main factor controlling the phase transformations in the temperature and pressure ranges studied is the concentration there alkali in the solution. With increasing MeOH concentrations there is a gradual exchange of the stable crystalline phases. In the $Na_2O-SiO_2-Al_2O_3-H_2O$ system we find analcite (albite), nepheline, cancrinite, sodalite; in the $K_2O-ZnO-GeO_2-H_2O$ system such phases include Zn_2GeO_4 and the KD phase; in the $Na_2O-GeO_2-H_2O$ system we have $GeO_2-Na_4Ge_9O_{20}$ and the H phase, and so on. These investigations are presented in full in [7, 16, 18].*

Conclusions
_____

1. In the hydrothermal crystallization of various substances, the number of compounds which are commonly used as the most effective solvents is quite limited. In the majority of cases crystals are grown on a seed under hydrothermal conditions in aqueous solutions of alkalis and the halides of monovalent metals.

2. Using specific examples, we have indicated the main requirements to be satisfied by the optimum solvent to be used in growing single crystals by the hydrothermal method.

3. We have distinguished various groups of solvents differing both as regards their properties and also as to the manner in which they behave under hydrothermal conditions.

_____

*The chemical formulas of the H and KD phases remain uncertain.

Literature Cited

1. S. Levinson and F. L. Carter, J. Electrochem. Soc., 13:756 (1966).
2. J. F. Balascio, R. B. White, and K. Roy, Mat. Res. Bull., 2:913 (1967).
3. L. N. Dem'yanets, L. S. Garashina, and B. N. Litvin, Kristallografiya, 8:800 (1963).
4. L. N. Dem'yanets, in: Hydrothermal Synthesis of Crystals [in Russian], Izd. Nauka (1968).
5. E. D. Kolb, A. S. Coriell, and R. A. Laudise, Mat. Res. Bull., 2:1099 (1967).
6. R. A. Laudise and A. A. Ballman, J. Phys. Chem., 65:1396 (1961).
7. I. P. Kuz'mina, Author's abstract [in Russian], IKAN SSSR, Moscow (1968).
8. V.A. Kuznetsov, Author's abstract [in Russian], IKAN SSSR, Moscow (1967).
9. R. A. Laudise and A. A. Ballman, J. Amer. Chem. Soc., 80:11 (1958).
10. I. P. Kuz'mina, Geologiya Rudnykh Mestorozhdenii, No. 3, p. 101 (1963).
11. J. Koenig, Solid State Phys., 12:210 (1961).
12. R. A. Laudise, J. H. Crocket, and A. A. Ballman, J. Phys. Chem., 65:359 (1961).
13. N. Yu. Ikornikova, V. A. Shorygin, and I. A. Vasil'chikova, in Growth of Crystals [in Russian], Vol. 4, Izd. Nauka (1964).
14. L. Yu. Khar'chenko, Author's abstract [in Russian], INKh SO AN SSSR, Novosibirsk (1968).
15. J. C. M. Henning, J. Liebertz, and R. P. Van Stapele, J. Phys. Chem. Solids, 28:1109 (1967).
16. B. N. Litvin and L. N. Dem'yanets, Kristallografiya, 7:643 (1962).
17. B. N. Litvin and L. N. Dem'yanets, Kristallografiya, 6:799 (1961).
18. O. K. Mel'nikov, Author's abstract [in Russian], IKAN SSSR, Moscow (1968).
19. B. N. Litvin, Author's abstract [in Russian], IKAN SSSR, Moscow (1964).
20. E. D. Kolb, D. A. Wood, E. G. Spencer, and R. A. Laudise, J. Appl. Phys., 38:1027 (1967).
21. B. N. Mill', Author's abstract [in Russian], IKAN SSSR, Moscow (1965).
22. R. A. Laudise and A. A. Ballman, J. Phys. Chem., 64:688 (1960).
23. I. P. Kuz'mina, Geologiya Rudnykh Mestorozhdenii, No. 1, p. 61 (1961).
24. W. Barnard, Econ. Geol., 62:138 (1967).
25. I. N. Anikin and A. D. Shushkanov, Kristallografiya, 8:128 (1963).
26. A. A. Ballman and R. A. Laudise, Hydrothermal Crystal Growth, The Art and Science of Crystal Growth (J. J. Gilman, ed.) Wiley, New York (1962), pp. 150-222.
27. G. W. Morey and E. Ingersol, Econ. Geol., 32:607 (1937).
28. W. Eitel, Physical Chemistry of Silicates [Russian translation], IL (1962).
29. R. Roy and O. Tattle, in: Physics and Chemistry of the Earth [Russian translation], IL (1958).
30. I. N. Anikin, Kritallografiya, 2:195 (1957).
31. E. N. Emel'yanova, S. V. Grum-Grzhimailo, O. N. Boksha, and T. M. Varina, Kristallografiya, 10:59 (1965).
32. I. E. Voskresenskaya, Author's abstract [in Russian], Moscow State University, Moscow (1968).
33. N. B. Keevil, J. Amer. Chem. Soc., 64:841 (1942).
34. G. W. Morey and W. T. Chen, J. Amer. Chem. Soc., 78:4249 (1956).

35. M. I. Ravich and F. E. Borovaya, Izv. Sekt. Fiz.-Khim. Analiza, 20:165 (1950).
36. S. Sourirajan and G. C. Kennedy, Amer. J. Sci., 260:115 (1962).
37. M. I. Ravich and F. E. Borovaya, Dokl. Akad. Nauk SSSR, 156:4 (1964).
38. V. I. Spitsyn and I. A. Savich, Zh. Obshch. Khim., 22:1278 (1952).
39. S. S. Chin, Zh. Fiz. Khim., 26:960 (1952).
40. S. S. Chin, Zh. Fiz. Khim., 26:1225 (1952).
41. A. I. Ellis and W. S. Fyfe, Rev. Pure and Appl. Chem., 4:261 (1957).
42. O. V. Bryzgalin, Geokhimiya, No. 6, p. 524 (1960).
43. V. S. Myasnikov, Dokl. Akad. Nauk SSSR, 33:412 (1941).
44. A. A. Beus, B. P. Sobolev, and Yu. D. Dikov, Geokhimiya, No. 3, p. 297 (1963).
45. L. H. Adams, J. Amer. Chem. Soc., 53:3769 (1931).
46. R. E. Gibson, J. Amer. Chem. Soc., 56:863 (1934).
47. R. E. Gibson, Amer. J. Sci., 235:49 (1938).
48. I. I. Friedman, Amer. Mineral., 34:583 (1949).
49. R. A. Laudise, J. Amer. Chem. Soc., 81:562 (1959).
50. R. A. Laudise and E. D. Kolb, Amer. Mineral., 48:5-6 (1963).
51. V. A. Shorygin, Kristallografiya, 8:5 (1963).
52. R. A. Laudise, E. D. Kolb, and A. J. Caporaso, J. Amer. Ceram. Soc., 47:9 (1964).
53. L. N. Dem'yanets, E. N. Emel'yanova, and O. K. Mel'nikov, Kristallografiya, 13:913 (1968).
54. R. L. Barns, R. A. Laudise, and R. M. Shields, J. Phys. Chem., 67:835 (1963).
55. H. Helgeson, Complexing in Hydrothermal Solutions [Russian translation], Izd. Mir (1967).
56. C. B. Monk, Electrolytic Dissociation, Academic Press, London (1961).
57. F. Rossotti, in: Modern Chemistry of Coordination Compounds (J. Lewis and R. Wilkins, eds.) [Russian translation], IL (1963).
58. F. Rossotti and H. Rossotti, Determination of the Stability Constants and Other Equilibrium Constants in Solutions [Russian translation], Izd. Mir. (1963).
59. C. J. Smithelles, Trans. Farad. Soc., 17:485 (1921).
60. H. Schafer, Chemical transport Reactions (N. B. Lezhnaya, ed.) [Russian translation], Izd. Mir (1964).
61. W. A. Chupka, J. Berkowitz, and C. F. Giese, J. Chem. Phys., 30:827 (1959).
62. A. Neubaus, Chim. Ing. Techn., 28:350 (1956).
63. V. Moozebakh, in: Thermodynamic of Geochemical Processes (V. V. Shcherbina, ed.) [Russian translation], IL (1960).
64. Ya. A. Fialkov and V. B. Spivakovskii, Zh. Neorg. Khim., 4, 1501 (1959).
65. V. A. Tyagai, V. S. Galinker, and G. N. Fenerli, Zh. Neorg. Khim., 7:1154 (1962).
66. V. V. Mironov, Zh. Neorg. Khim., 8:3 (1963).
67. V. B. Mironov, F. Ya. Kul'ba, and V. A. Nazarov, Zh. Neorg. Khim., 8:916 (1963).
68. F. T. Smit, in: Questions of Physical Chemistry in Mineralogy and Petrography [Russian translation] (D. S. Belyankin, ed.), IL (1950).
69. N. I. Khimarov and T. N. Kozintseva, in: Experimental Studies in the Field of Subsurface Processess (N. I. Khitarov, ed.) [in Russian], Izd. AN SSSR (1962).
70. B. Mason, J. Geol., 57:1 (1949).

71. M. Pourbaix, Atlas d'Équilibres Électrochimique, Paris (1963).
72. C. A. Chapman and G. K. Schweitzer, J. Geol., 55:43 (1947).
73. M. M. Jones, J. Amer. Chem. Soc., 76:4233 (1954).
74. Yu. V. Morachevskii and L. I. Lebedeva, Zh. Neorg. Khim., 5:10 (1960).
75. J. Jander, Z. Anorg. Chem., 194:382 (1930).
76. H. Brintzinger and W. Brintzinger, Z. Anorg. Chem., 196:58 (1931).
77. R. G. Yalman, E. R. Shaw, and J. F. Corwin, J. Phys. Chem., 64:300 (1960).
78. G. N. Malcon, M. N. Parton, and I. D. Watson, J. Phys. Chem., 65:1900 (1961).
79. M. A. Styrikovich, in: Thermodynamics and Structure of Solutions (M. I. Shakhparanov, ed.) [in Russian], Izd. AN SSSR (1959).

Synthesis of Zincite by the Hydrothermal Method

I. P. Kuz'mina, A. N. Lobachev, and N. S. Triodina

Zincite crystals may be grown in a variety of ways: from the gas phase, yielding acicular crystals [1], by the crystallization of zinc oxide from solution in PbF_2 [2], which gives imperfect, lamellar crystals with (0001) and (000$\bar{1}$) basic faces, and so on. The properties of crystals grown by these methods are technologically unsatisfactory.

The most promising method of producing isometric zinc oxide crystals of good quality is the hydrothermal method. Zincite crystals of practical value may be obtained in this way. The fairly high growth rates and the continuous, reproducible growth process on a seed ensure the production of crystals with dimensions which are simply determined by the area of the seed and the period of the experiment [3].

In order to obtain zincite by the hydrothermal method, the solvents usually employed are aqueous solutions of NaOH or KOH with a strength 4-6 M [3]. The temperature of the dissolution zone is 380-410°C, that of the growth zone 350-380°C, and the temperature drop 10-30°C. The pressure is determined by the occupation factor of the autoclave, which is between 78 and 88%. As initial charge, zinc oxide chemical reagent or a fine-crystalline charge of ZnO previously recrystallized under hydrothermal conditions is employed. As mineralizers for improving the growth conditions, lithium salts (LiOH, LiF) of concentration 0.1-2 M, are usually added to the alkali solutions.

27

The use of alkalis sometimes leads to the formation of hydrogen, arising as a result of the oxidation of the autoclave material. The hydrogen pressure in the autoclave reaches 60-80 atm. The amount of hydrogen formed depends greatly on the temperature, the concentration of the alkali, and the duration of the experiment.

It should be noted that the hydrogen formed during the growth of the zincite crystals has a considerable effect on the properties of the crystals, worsening their quality. The dissolution of the crystals, the transport processes, and crystallization in turn depend greatly on the redox potential of the medium [4].

Zincite crystals obtained under such conditions have a low electrical resistance ($\sim 10 \ \Omega \cdot cm$), this probably being due to the disruption of their chemical composition. In order to increase the resistance of the crystals the latter are further annealed in air or oxygen.

The zincite crystals as grown have the following simple shapes: hexagonal prisms ($10\bar{1}0$), hexagonal pyramids ($10\bar{1}1$), and two monohedra (0001) and ($000\bar{1}$). The growth rate of the (0001) face is approximately 8-10 times greater than that of the prism and pyramid faces. The growth rate of the prism ($10\bar{1}0$) faces is 10-15% higher than that of the pyramid ($10\bar{1}1$) faces, so that in the course of growth the slowly-growing pyramid faces appear on the zincite crystal, the (0001) face of the monohedron tapers off, and further growth of the crystal is substantially retarded. What is required to improve the situation is a ratio of the growth rates of the prism and pyramid faces such as would increase the pyramid growth rate. In that case only the faces of the prism and monohedra would be present on the crystal, and the crystal would grow at the same rate of an almost unlimited time, the latter being simply determined by the duration of the experiment.

We found in our experiments that the ratio of the prism and pyramid face growth rates depended on the temperature drop in the system. In addition to this, earlier published data [5] and our own more recent experimental results showed that the presence of various impurities in the solution had a major effect on the ratio between the growth rates of the various faces.

After the zincite crystals had been grown, they were studied by micro- and macrocrystallographic techniques in order to discover the general growth laws. The relief of the monohedral (0001) and (000$\bar{1}$) faces, the prism face (10$\bar{1}$0), and the pyramid (10$\bar{1}$1) face was considered in this connection.

The composition of the solution in which the zinc oxide crystallizes changes the habit and properties of the developing crystals. Thus, in KOH solutions, ZnO crystals or prismatic habit, drawn out sharply along the c axis, are formed; in caustic soda solutions the crystals are more isometric, in LiOH solutions the growth rate of the monohedral face is very slow, while that of the pyramid face is commensurable with or slightly exceeds that of the prism, and the crystals assume a tabloid form. Hence in order to obtain good crystals growth is usually carried out in NaOH and KOH solutions.

We found that the solution used in crystallization also (to a certain extent) determined the quality of the developing crystals. Thus in NaOH solutions the crystals are less perfect than in KOH. The resistance of crystals obtained in these solutions is of roughly the same order, being equal to a few ohm-centimeters.

The growth rates v depend on the supersaturation ΔS of the solution, any change in which leads to a change in the facing of the crystal. By measuring the growth rates of the faces in mm of grown layer per unit time (days) we studied the v = f (ΔT) relationship for the (0001) and (000$\bar{1}$) faces, and in individual cases for the (10$\bar{1}$0) faces as well. The resultant data are presented in Table 1, which

TABLE 1

Serial No.	ΔT, °C	ΔS wt.%	v (0001)	v (000$\bar{1}$)	v (10$\bar{1}$0)
			mm/day		
1 *	5	0.034	0.082		
2	9	0.0612	0.17	0.028	
3 *	12	0.086	0.259		
4	12	0.0816	0.20	0.10	0.05
5	14	0.0952	0.27	0.14	
6 **	14	0.0952	0.27	0	
7 *	15	0.102	0.286		
8	15	0.102	0.271	0.10	0.0475
9	15	0.102	0.31	0.14	0.045
10 *	20	0.136	0.40		

* Results of [2].
** In the presence of KClO$_3$.

also contains data relating to the (0001) face [6]. On the basis of the measurements so carried out we plotted the graphical relationships $v_{(0001)} = f(\Delta T)$, $v_{(000\bar{1})} = f(\Delta T)$ shown in Fig. 1.

We see from Fig. 1 that the growth rates of the two faces increase linearly with the relative supersaturation of the solution. By extending the resultant straight lines to intersect the horizontal axis, we obtain the points of critical supersaturation for the two faces: $\Delta T_{cr} \approx 1°C$ and $\Delta T_{cr} \approx 8°C$. These quantities represent the critical saturations below which the growth of the two monohedra fails to occur. This effect is observed in actual practice; for small temperature drops the growth of the (0001) face predominates (in the direction of the +c axis); the growth of the other (000$\bar{1}$) face in the −c direction is negligible. The experimental value of $v_{(10\bar{1}0)} = 0.045$ mm/day for $\Delta T = 15°C$ suggests that the ΔT_{cr} for the prism face (10$\bar{1}$0) equals 12–14°C. The calculated values of the critical supersaturations equals 0.0068 wt.% for the (0001) face and 0.0545 wt.% for the (000$\bar{1}$) face. Thus we have the relationship $\Delta T_{cr(0001)} < \Delta T_{cr(000\bar{1})} < \Delta T_{cr(10\bar{1}0)}$, which is the reverse of the velocity relationship $v_{(0001)} > v_{(000\bar{1})} > v_{(10\bar{1}0)}$. The $v_{(0001)}/v_{(000\bar{1})} = f(\Delta T)$ curve is illustrated in Fig. 2.

The necessity of producing zincite crystals possessing a high specific resistance has compelled many research workers to grow these crystals in the presence of various impurities.

Published data show [6] that one of the most important mineralizers is the lithium ion, which plays a multitude of roles.

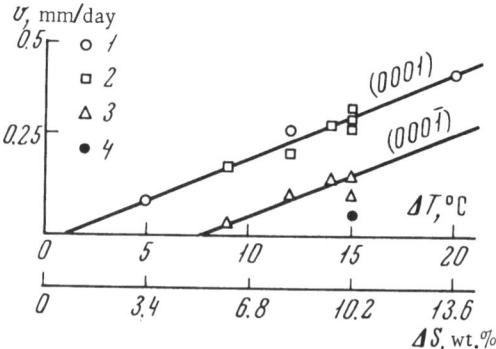

Fig. 1. Dependence of the growth rates v of the (0001) and (000$\bar{1}$) faces on the supersaturation ΔT (ΔS). 1) $v_{(0001)}$, from [5]; 2) $v_{(0001)}$; 3) $v_{(000\bar{1})}$; 4) $v_{(10\bar{1}0)}$.

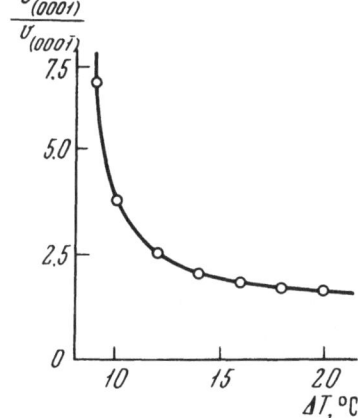

Fig. 2. Dependence of the growth-rate ratio $v_{(0001)}/v_{(000\bar{1})}$ on the relative supersaturation ΔT.

The presence of lithium in the solution affects both the crystalliza-tion processes and the properties (mechanical and physical) of the crystals so formed. Usually lithium is added to the solution in the form of LiOH, $LiBO_2$, and LiF in quantities of 1.0-2.0 M. In the presence of lithium, relatively perfect zincite crystals, with-out any signs of cracking, are obtained. The surface of the (0001) face becomes smooth and even in the presence of lithium. This effect was explained in [5] as being due to the adsorption of lithium on the (0001) face of the monohedron and thus to a reduction in its free surface energy, leading to a fall in the growth rate of the (0001) face.

At the same time [6], the introduction of lithium into the solution slightly increases the growth rate of the prism faces ($10\bar{1}0$). Unfortunately only qualitative results were given in [6], without any quantitative measurements of the growth rates of the various faces of the ZnO crystals.

The resistance of zincite crystals grown in alkali solutions containing lithium ions is small: 10^1-10^2 $\Omega \cdot cm$. Annealing in air or oxygen increases the resistance of the samples to 10^5-$10^8 \Omega \cdot cm$. This is due to the diffusion of the lithium ions, and to the motion of the hyperstoichiometric proportion of zinc to the surface of the crystal faces, with its subsequent oxidation to zinc oxide. The highest values of the resistance (10^9-$10^{12} \Omega \cdot cm$) are obtained for zinc oxide samples grown in the presence of 2 M LiOH [6].

Among the impurities which affect not only the properties of the developing crystals but also the kinetics of zincite crystallization is the NH_4^+ ion. The presence of this ion in solution during crystallization increases the rate of growth of the prism faces by a factor of approximately five times, while the growth rates of the other faces remain the same as before [5]. The mechanical properties of the developing crystals, however, become slightly less favorable; microscopic cracks and other defects appear.

We found that zincite crystals were excellently activated by manganese and nickel, which respectively colored them red and green. The resistance of the crystal remained unchanged. Manganese and nickel ions also had no appreciable effect on the kinetics of the crystallization of zincite. The presence of F' ions in solution increases the growth rate of the monohedral (0001) face, slightly (by an order of magnitude) increasing the resistance of the developing crystals. However, if the solution contains iron, characteristic zincite twins growing together along the face of the monohedron are formed. The resultant crystals contain a large quantity of iron (up to several percent). It should be noted that the formation of zincite twins and triplets has also been observed on crystallizing zincite in solutions of ammonium chloride.

An oxidizing anneal in an oxygen atmosphere in order to obtain a higher resistance of the grown crystals is not always appropriate, since it requires long experiments at fairly high temperatures and does not always produce zincite crystals with stable properties; the resistance varies from sample to sample. It is therefore better to look for zincite growth conditions which will produce samples of high resistance directly.

The fact that the stoichiometric compositions of zincite crystals is disrupted under reducing conditions of crystallization suggests that the crystals ought rather to be grown in the presence of oxidizing agents. We studied the crystallization of zincite in solutions containing $KClO_3$, $KMnO_4$, Br_2, etc. The resistance of the crystals grown under these conditions increased by two or three orders of magnitude, i.e., the introduction of oxygen into the crystallization medium made a great difference to the properties of these crystals. The introduction of the oxidizing agent into the solution also affects the crystallization kinetics, reducing the rate of growth of all faces of the crystal, including the fastest-growing monohedral (0001) face. The growth of the ($000\bar{1}$) face in the -c direction, however, ceases altogether, the crystal only growing in the +c direction. Thus the introduction of an oxidizing agent improves the properties of the growing crystals but has a serious effect on the crystallization kinetics, retarding the growth of all faces. This is evidently associated with the fact that the impurities are adsorbed on the various faces and change their growth rate accordingly.

In order to understand all the processes taking place during crystallization, micromorphological analyses of the zincite faces are essential. We carried out an analysis of this kind using the MIM-8M metallographic microscope. Let us now consider the results.

For the (0001) monohedral face a characteristic feature is a smooth, specular surface, often containing hexagonal microchannels, the number of which changes with the experimental conditions (Fig. 3). We may suppose that these channels originate from the deposition of very fine particles of the original charge (suspended in the solution) on the face of the crystal; on subsequent growth of the latter these distort the relief of the face. Since the rate of the growth of the (0001) face is greater than that of the prismatic ($10\bar{1}0$) face and also greater than that of the pyramid ($10\bar{1}1$), the prismatic face grows considerably more slowly, and is thus unable to "grow around" the particle of the charge; a growing crater therefore appears in the crystal. At the bottom of the crater is an aggregate of charge particles, forming the origin of a hollow, non-overgrown channel on the (0001) face of the crystal (Fig. 4).

Fig. 3. Hexagonal macrochannels on the (0001) face (× 100).

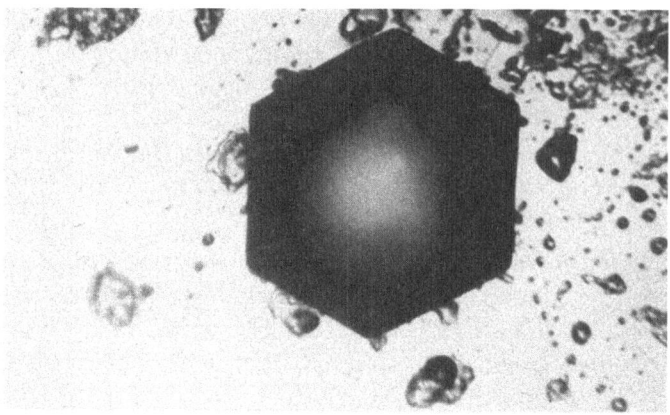

Fig. 4. Aggregation of charge particles at the botton of a hexagonal channel (× 200).

Under growth conditions characterized by high supersaturations, unfilled hollow channels are usually not formed. Nor do
these appear when the original charge comprises zinc oxide previously recrystallized under hydrothermal conditions.

With increasing supersaturation a particle becoming attached
to the seed becomes a growth center in its own right. Figure 5

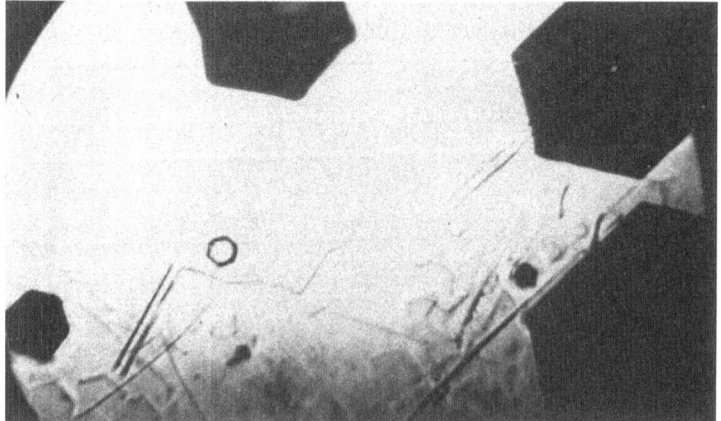

Fig. 5. Overgrowth of hexagonal apertures on the (0001) face with
increasing supersaturation (× 100).

Fig. 6. Spiral hexagonal growth pyramids (× 400).

clearly shows both partly and completely overgrown openings.
As the duration of the experiment increases (the coarseness of the
charge increasing at the same time on account of recrystallization
in the hot zone of the autoclave), craters no longer form on the
(0001) face, and two characteristic regions are revealed. One region
constitutes a mirror-smooth surface, on which detailed examination
(with a phase-contrast adapter) reveals spiral hexagonal growth
pyramids (Fig. 6). The other region on this face is characterized
by conical layer-like growth accessories (Fig. 7).

The relief of the (000$\bar{1}$) face is characterized by the presence
of disc-like growth figures of different heights (Fig. 8). In indiv-

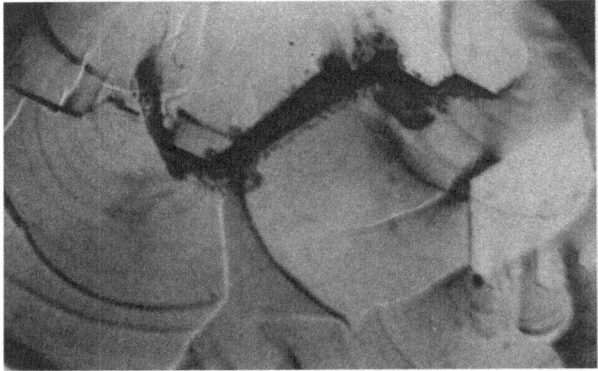

Fig. 7. Conical layer-like growth accessories (× 100).

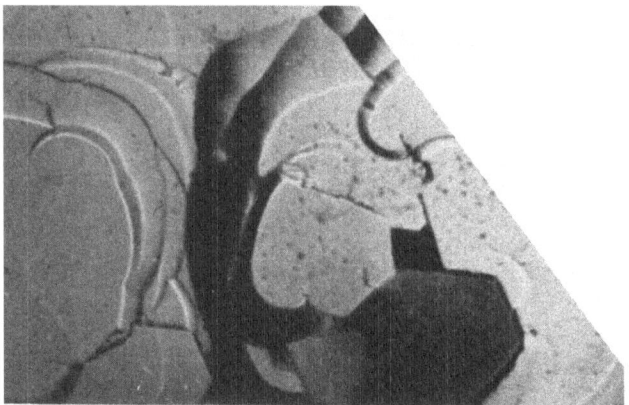

Fig. 8. Disc-shaped growth figures on the (000$\bar{1}$) face (× 100).

idual regions there are layer-like growth accessories of bizarre
form (Fig. 9). A change in the relative supersaturation has a
specific influence on the microrelief of the surface. The face with
the least supersaturation is covered with conical growth accessories
having flat hexagonal vertices sharply reflecting the crystal symmetry (Fig. 10). An increase in supersaturation (Fig. 11) leads
to an increase in the dimensions of the hexahedral vertices of the
accessories, which grow together for very high supersaturations
and form whole regions with a specular surface (Fig. 12).

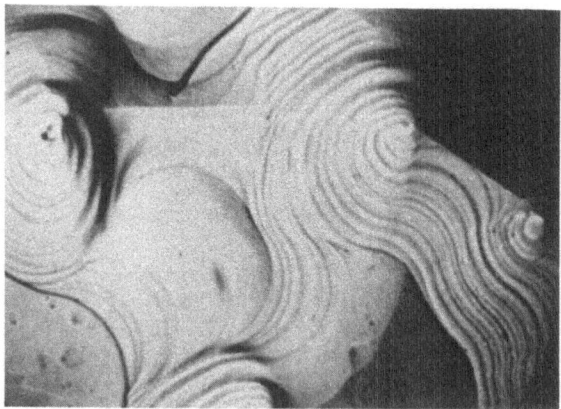

Fig. 9. Layer-like growth accessories (× 100).

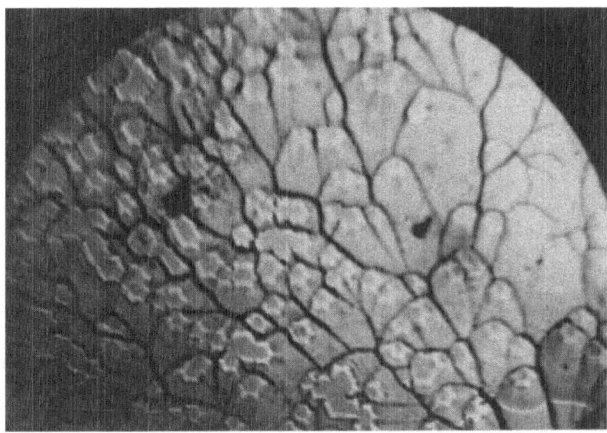

Fig. 10. Layer-like conical growth accesories with flat
hexagonal vertices (× 100).

A more detailed examination of the surface of the (000$\bar{1}$) face
at the lowest supersaturations reveals spiral-layer growth figures (Fig. 13).

The (10$\bar{1}$0) prism face develops in the main as a result of
layer-like growth (Fig. 14). No changes occur in the character of
the relief on changing the relative supersaturation.

On the (10$\bar{1}$1) pyramid face, growth regions of various charac-
ters appear for large supersaturations. In one of these, a hilly re-

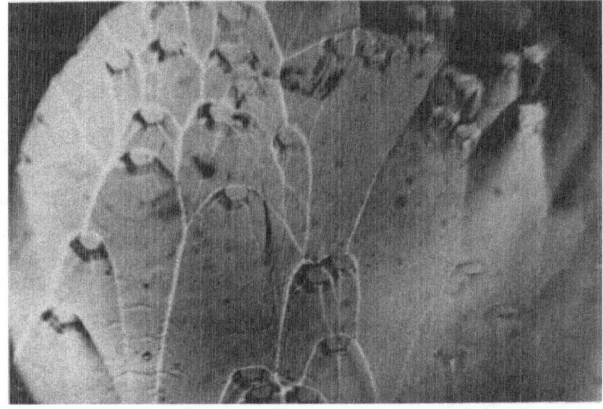

Fig. 11. Increase in the dimensions of the hexagonal vertices of the accessories with increasing supersaturation (× 100).

Fig. 12. Formation of regions with a specular surface on
the $(000\bar{1})$ face with increasing supersaturation ($\times$ 100).

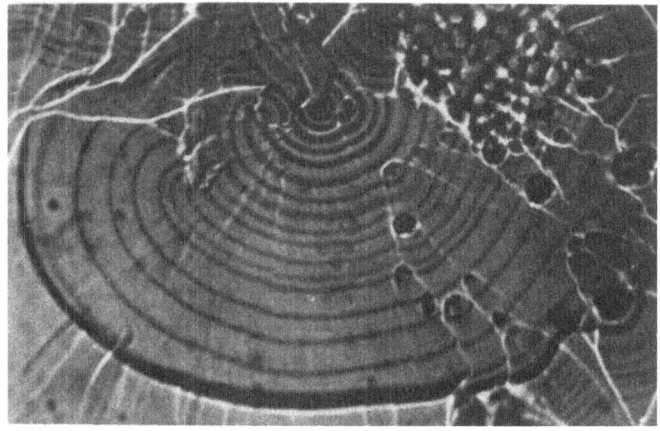

Fig. 13. Spiral-layer growth figures ($\times$ 100).

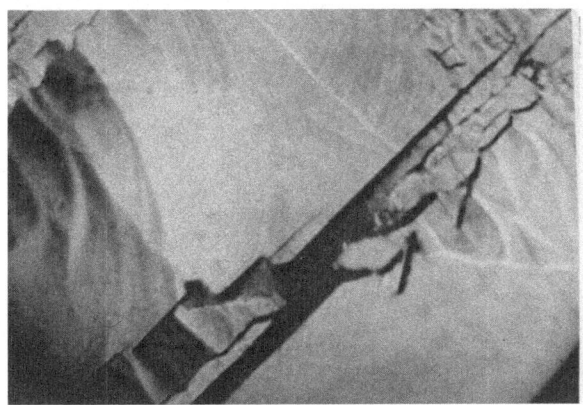

Fig. 14. Characteristic relief of the (10$\bar{1}$0) prism face
($\times$ 100).

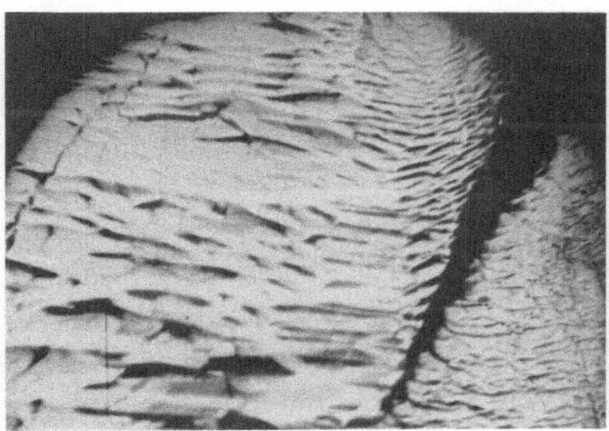

Fig. 15. Hilly surface of the (10$\bar{1}$1) pyramid face ($\times$ 100).

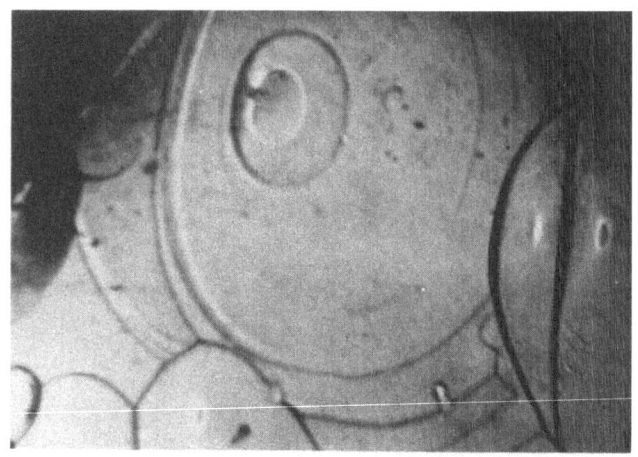

Fig. 16. Layer-like growth figures on the (10Ī1) pyramid
(× 100).

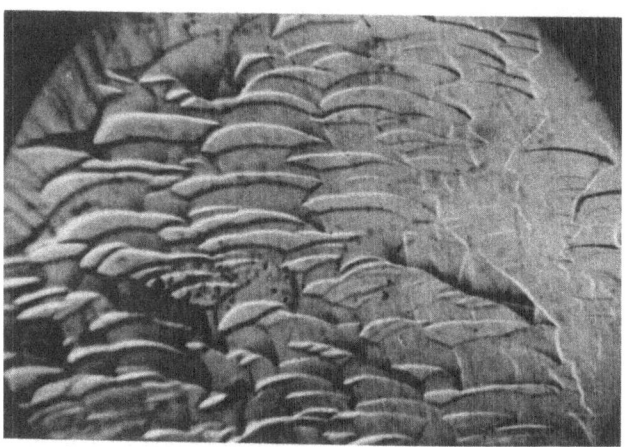

Fig. 17. Relief of the (10Ī1) face for slight supersaturations
(× 100).

lief develops (Fig. 15), while the rest of the surface is covered with layer-like growth figures (Fig. 16). For small supersaturations growth accessories appear on the faces, the slopes of these exhibiting terraces (Fig. 17).

Conclusions

1. We have studied the growth rates of various zincite crystal faces as functions of the degree of supersaturation, and established the critical supersaturations for the (0001) and (000$\bar{1}$) faces.

2. We have determined the role of a number of impurities in the crystallization of zincite under hydrothermal conditions. We have determined the effect of impurities on the properties of the developing crystals. We have also studied the crystallization of zincite in the presence of an oxidizing agent.

3. We have carried out a morphological analysis of the faces of zincite crystals.

Literature Cited

1. I. Kulo, Japan. J. Appl. Phys., 4:225 (1965).
2. I. W. Nielsen and E. F. Dearborn, J. Phys. Chem., 64:1762 (1960).
3. R. A. Laudise and A. A. Ballman, J. Phys. Chem., 64:688 (1960).
4. B. I. Mason, in: Questions of Physical Chemistry in Mineralogy and Petrography [Russian translation] (D. S. Belyankin, ed.), IL (1950).
5. E. D. Kolb and R. A. Laudise, J. Amer. Ceram. Soc., 49:303 (1966).
6. R. A. Laudise, E. D. Kolb, and A. J. Caporaso, J. Amer. Ceram. Soc., 47:9 (1964).

Crystallization of the Oxides of Titanium Subgroup Metals

V. A. Kuznetsov

In this article we shall generalize a number of investigations which have been made into the crystallization of titanium, zirconium, and hafnium oxides under hydrothermal conditions. There are two main reasons for carrying out such investigations. Firstly, the interest currently expressed by physicists in these compounds calls for constant research into methods of producing high-quality single crystals. The high melting points of ZrO_2 and HfO_2 (2700 and 2780°C [1]) makes it difficult to grow the corresponding crystals directly from the melt. The crystallization of rutile by the Verneuil method is frequently employed, but unfortunately this method often yields crystals of a nonstoichiometric nature. Attempts have accordingly been made at one time and another to grow zirconium, hafnium, and titanium oxides at lower temperatures by using various fluxes.

Secondly, experiments on growing various titanates and zirconates of divalent metals by the hydrothermal method showed that titanium and zirconium constituted "inert," low-mobility components, preventing the recrystallization of the complex oxides of which they formed a part. Clearly any explanation which might be offered for the transfer behavior of TiO_2 and ZrO_2 would assist in solving the problem of the synthesis and recrystallization of titanates and zirconates.

Among the various possible methods of growing TiO_2, ZrO_2, and HfO_2 single crystals (apart from the well-known and widely-

used Verneuil method of growing rutile), the most widely-accepted
techniques are those based on crystallization from solution in
molten salts. In the case of ZrO_2 and HfO_2 the fluxes used have included Li_2MoO_4 [2], borax [3], and PbF_2 [4]. The latter flux gave
the best results, lamellar ZrO_2 and HfO_2 crystals of up to 6 and 3
mm in size respectively being obtained. In growing TiO_2, apart
from the salts already mentioned, use has also been made of
Na_3AlF_6, Na_2WO_4, Na_3PO_4, fluorides, carbonates, alkalis, etc. [5-8].
The crystallization of rutile from the gas phase was mentioned
in [8] as well. In this case the titanium was transported in the
form of halides, such as $TiCl_4$ and TiF_4.

We are not aware of any experimental work relating to the
crystallization of zirconium and hafnium oxides under hydrothermal conditions. The hydrothermal crystallization of TiO_2, however,
was considered by I. N. Anikin et al. [8] at temperatures above
550°C and pressures of over 1000 atm, and M. L. Harville and R.
Roy [9] in H_2SO_4 solutions at 700°C and 1000-4000 atm. However,
the crystal growth rate in these experiments was very low.

In the present investigation we considered the behavior of
a large group of solvents in order to select the most promising of
these for the recrystallization of titanium, zirconium, and hafnium oxides. The solvents considered included solutions of alkalis,
lithium, potassium, sodium, and ammonium carbonates and bicarbonates, sodium sulfide, lithium, potassium, and ammonium chlorides,
sulfates (Na_2SO_4, K_2SO_4), borax, boric acid, and fluorides (KF,
NaF, NH_4F). The concentrations of the solutions were 5-40%.*
The main results of this exploratory part of the work may be sum-
marized as follows.

In alkali solutions and solutions of potassium, sodium, and
lithium carbonates and bicarbonates, the titanates, zirconates, and
hafnates of the corresponding metals were formed. In the case of
TiO_2, titanates were synthesized in the form of thin filamentary
crystals up to 2-3 mm long for all concentrations of the original
solutions (5-40%). At the same time, titanates of a different chemical
composition but very similar external habit were formed. This
made it difficult to select individual phases for subsequent diagnosis
by x-ray and chemical methods. The formation of zirconates and

---

*All the work was carried out in standard autoclaves 150 cm³ in volume (up to 600°C)
and in an exoclave made by A. A. Shternberg (above 600°C). The pressure was varied
from 800 to 3000 atm.

TABLE 1

Charge, g	Dissolution temperature, °C	Growth temperature, °C	Solvents (10 wt.%)	Duration of experiment, days	Weight of transported material, g	Dimensions of crystals, mm
TiO$_2$ 20	500	470	NaF	6	0.5	0.3
TiO$_2$ 20	600	570	NaF	6	3.0	0.8
TiO$_2$ 20	500	470	KF	6	3.0	0.3
TiO$_2$ 20	600	570	KF	6	14.0	1.5
TiO$_2$ 20	470	500	NH$_4$F	6	Whole charge transported	1.0
ZrO$_2$ 20	550	520	NaF	6	No transport	
ZrO$_2$ 3	700	690	NaF	2	0.2	0.3
ZrO$_2$ 20	550	520	KF	6	No transport	
ZrO$_2$ 20	600	570	KF	5	0.3	0.3
ZrO$_2$ 3	700	690	KF	2	0.5	0.5
ZrO$_2$ 20	520	550	NH$_4$F	6	Whole charge transported	0.5
HfO$_2$ 20	600	570	NaF, KF	6	No transport	
HfO$_2$ 2	700	690	NaF, KF	2	0.2	0.2
HfO$_2$ 20	520	550	NH$_4$F	6	Whole charge transported	1.0

hafnates (in the form of acicular crystals up to 1.5 mm long) started at higher concentrations of the solutions than those corresponding to the synthesis of titanates; in alkali solutions at concentrations of over 15%, in carbonates over 25%, and in bicarbonates over 35%. For lower concentrations of the solutions the original charge recrystallized, and ZrO$_2$ and HfO$_2$ crystals up to 0.3 mm long formed in the lower part of the autoclaves (at 600°C). No substantial transfer of ZrO$_2$ and HfO$_2$ to the upper zone of the autoclave occurred in alkali, carbonate, and bicarbonate solutions.

In chloride, sulfate, sulfide and borate solutions only recrystallization of the original material *in situ* took place under the conditions in question, without any substantial transport of material to the upper zone of the autoclave.

The titanium, zirconium, and hafnium dioxides dissolved and recrystallized with the greatest intensity in solutions of KF, NaF, and NH$_4$F. We studied these solutions in greatest detail. The high mobility of titanium, zirconium, and hafnium in these solutions enabled us to grow TiO$_2$, ZrO$_2$, and HfO$_2$ crystals by the temperature-drop method. Table 1 shows the results of some experiments relating to the spontaneous crystallization of these oxides in 10% solutions of fluorides.

We should note two points characterizing the behavior of titanium, zirconium oxides in fluoride solutions.

(a) The optimum concentrations of the solutions for the recrystallization of the oxides in 7-10%. In more concentrated solutions, crystals consisting of fluoride compounds of titanium, zirconium, and hafnium are formed. We were unable to identify these compounds unambiguously by reference to the x-ray diffraction patterns, but spectral analysis confirmed the presence of a large quantity of fluorine, and in a number of cases sodium or potassium. The formation of the fluoride compounds has a bad effect on the recrystallization of TiO_2, ZrO_2, and HfO_2; the transport of the oxides diminishes and then ceases altogether.

(b) The most important distinguishing feature of fluoride solutions is the fact that the direction of mass transfer in these depends on the solvent cation; under all conditions studied in KF and NaF solutions the recrystallization of TiO_2, ZrO_2, and HfO_2 proceeded from the "hot" to the "cold" zone of the autoclave, and in NH_4F solutions it porceeded in the opposite direction. This characteristic, as we shall subsequently see, may be attributed to the different stabilities of sodium and potassium fluorozirconates, fluorotitanates, and fluorohafnates, on the one hand, and the corresponding ammonium salts on the other, in high-temperature solutions.

Crystallization of TiO_2 (Rutile) from Fluoride Solutions

In 7-10% solutions of KF or NaF a fairly intensive recrystallization of TiO_2 starts at temperatures of over 500-550°C and pressures of 500-800 atm. The amount of material transported increases with increasing temperature of the experiment, while for equal temperatures and pressures it is considerably greater in KF than in NaF solutions. We therefore mainly used KF solutions in the spontaneous crystallization of rutile and the growth of rutile on a seed.

Figure 1 shows TiO_2 crystals obtained by spontaneous crystallization in 10% solutions of KF at 550°C. The dimensions of the crystals may extend to as much as 1-1.5 mm in experiments lasting for three days. Increasing the crystallization temperature usually makes the crystals larger. In a special series of exper-

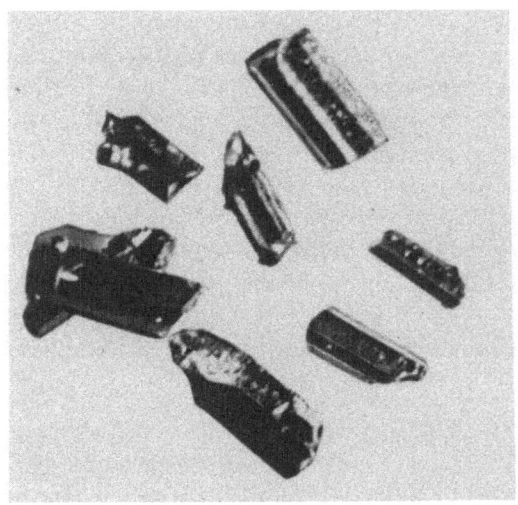

Fig. 1. Hydrothermal rutile crystals (× 10).

iments, a rutile crystal grown by the Verneuil method was placed in the upper zone of the autoclave to act as seed (Fig. 2). Growth on the seed started at 450°C (~800 atm, $\Delta T = 30°$); however, the growth rate of the crystal faces was very slow under these conditions. With increasing temperature the rate of crystallization became faster, and at 550°C the growth rate of the (110) and (100) faces reached 0.2 and 0.3 mm/day respectively. In the ⟨001⟩ direction the growth rate was still higher.

Fig. 2. Growth of rutile on a seed (× 4).

As compared with KF and NaF solutions, solutions of NH_4F (5-10%) produce a more intensive recrystallization of TiO_2, starting at lower temperatures; even at 400-500°C, 20 g of charge is completely transferred in experiments lasting 5-6 days. However, in view of the fact that the material passes from the "cold" to the "hot" zone of the autoclave, the original reagent has to be placed in the upper zone of the reactor, which is inconvenient when using the normal experimental method (in vertical autoclaves). For this reason, as well as because of the fact that a perfectly acceptable crystal growth rate may be achieved in KF solutions, there is no great point in using NH_4F solutions for growing rutile crystals.

Our rutile crystal usually have a long- or short-prismatic habit (Fig. 1). The most widely-encountered forms are {110}, {100}, {111}, although {101}, {321} are often present as well. Articulated twins and triplets often appear along the (101), these being characteristic of natural rutile crystals, as well as other more complicated forms of contiguous growth.

The crystals are usually colored in dark tints and frequently almost black (Figs. 1 and 2). In a number of cases the coloring is due to traces of iron leached from the walls of the autoclave. However brownish rutile crystals have also been obtained in autoclaves lined with copper, silver, and platinum linings. In this case the color is associated with the deviation of the crystals from stoichiometric ratio.

At high oxygen pressures (achieved by introducing $KClO_3 \sim$ 0.5 g into the autoclave), light yellow TiO_2 crystals are obtained.

Crystallization of ZrO_2 and HfO_2 in Fluoride Solutions

The crystallization of ZrO_2 and HfO_2 in KF and NaF solutions starts at higher temperatures than the recrystallization of TiO_2 and is far less vigorous. Thus in KF solutions ZrO_2 is only transported at temperatures of over 600°C, and HfO_2 at over 650-670°C, but even at these temperatures the amount of material transported is very little; only individual crystals of ZrO_2 and HfO_2 up to 0.5 mm in size (Fig. 3) are formed in the upper zone of the autoclave. Below 600°C, all that happens in that the original material is made coarser, and lamellar crystals up to 0.3 mm in size are formed in the lower zone of the autoclave.

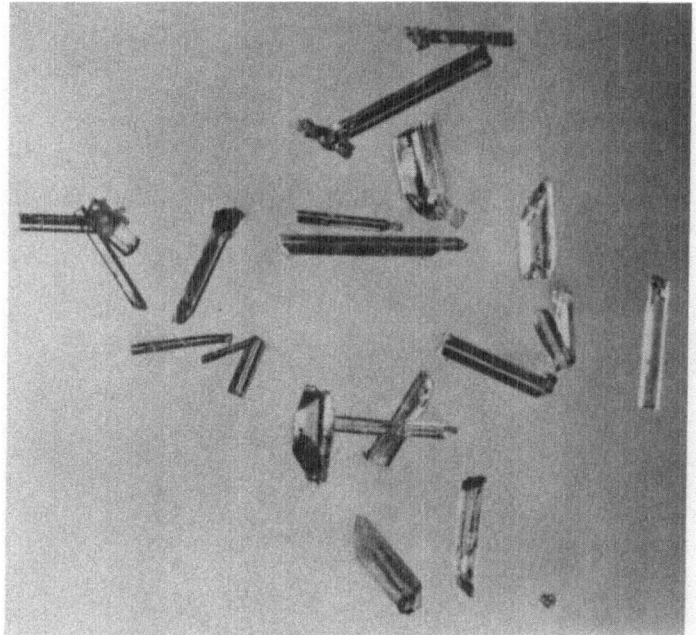

Fig. 3. ZrO_2 crystals obtained in KF solutions ($\times$ 10).

The situation is quite different in NH_4F solutions. In these there is an intensive recrystallization of ZrO_2 and HfO_2 even at relatively low temperatures, and at 500–550°C 20 g of charge is transported in 4–6 days. Druses of ZrO_2 and HfO_2 crystals are formed in the "hot" zone of the autoclave, the size of individual crystals reaching 2 mm.

Solutions of NH_4F are practically the only ones suitable for growing ZrO_2 and HfO_2 crystals by the temperature-drop method. However, the retrograde solubility of ZrO_2 and HfO_2 in the solutions creates a number of experimental difficulties. For example, in a vertical autoclave it is essential to place the original charge in the upper zone of the reactor and to effect recrystallization in the lower zone, from which it is difficult to extract the crystals without breaking them. In order to simplify the problem, we used a furnace with a horizontal autoclave and created a "hot" zone at the obturator. With this type of autoclave it was considerably easier to regulate the mass-transfer velocity (by varying the temperature drop) and extract the resultant crystals than with the vertical autoclave. Using the horizontal autoclave we obtained ZrO_2 crystals

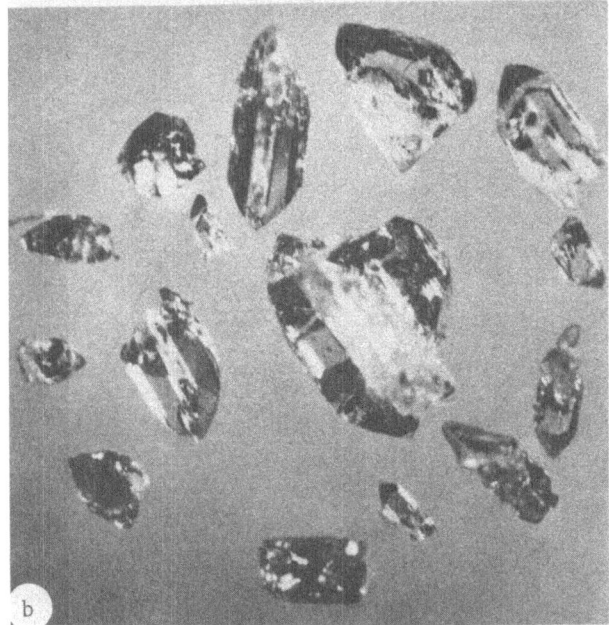

Fig. 4. Crystallization of ZrO_2 in a solution of NH_4F. a)
$\Delta T = 15°$ (× 6); b) $\Delta T = 30°$ (× 10).

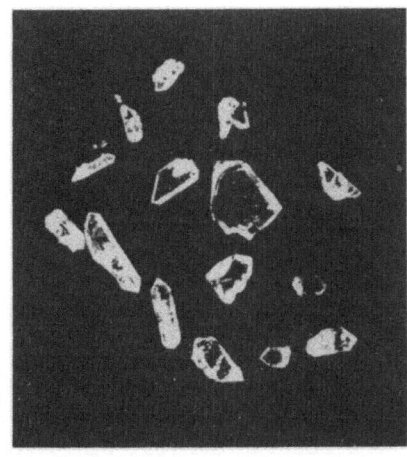

Fig. 5. Hydrothermal crystals of HfO$_2$
(× 5).

in the form of thin plates 2.5 × 2.5 × 0.5 mm in size (Fig. 4a). An increase in the temperature drop leads to a more intensive crystallization and tend to thicken the crystals; however, the crystals then undergo substantial twinning (Fig. 4b). We obtained HfO$_2$ crystals in an analogous manner (Fig. 5). Table 2 indicates some typical conditions for experiments in a furnace with a horizontal autoclave.

Since we were unable to obtain any large zirconium and hafnium oxide crystals we were unable to study the growth of these oxides on a seed and estimate the growth rates of the different faces. From the morphology of the spontaneously-formed crystals we may conclude that the maximum growth rate occurs in the c direction; growth is slightly slower in the b, and slowest in the *a* directions. The growth rate of the (100) face is roughly 10 times lower than that of the (001).

TABLE 2

Charge	Dissolution temperature, °C	Growth temperature, °C	Occupation factor	Duration of experiment, days	Size of crystals, mm
ZrO$_2$	550	560	0.55	7	1.0×1.0×0.3
ZrO$_2$	550	570	0.55	7	2.0×2.0×0.5
ZrO$_2$	550	585	0.5	7	2.5×2.5×0.8
ZrO$_2$	590	610	0.5	5	2.5×2.5×0.8
HfO$_2$	570	585	0.5	7	1.0×1.0×0.3
HfO$_2$	570	600	0.5	6	1.5×1.5×0.5

According to x-ray diffraction data, the resultant ZrO_2 and HfO_2 crystals belong to the monoclinic system; they are mainly represented by lamellar forms with predominantly (001), (010), and (110) faces. The ZrO_2 crystals often twin along the (110) or (101), forming characteristic cruciform interpenetrations. In contrast to TiO_2, the ZrO_2 and HfO_2 crystals are colorless and transparent; only in the presence of iron do they assume brownish tints.

Titanium, Zirconium, and Hafnium Transport Forms

In order to determine the manner in which the titanium, zirconium, and hafnium atoms are transported, it is important to take account of the very different ways in which the corresponding dioxides crystallize in fluoride and alkali solutions (titanates, zirconates, and hafnates being formed in the latter), as well as the capacity of titanium, zirconium, and hafnium to form strong complexes with fluorine. It follows quite logically from these properties that, in the presence of alkali metals, Ti, Zr, and Hf cannot be transported in the form of oxygen-containing anions; transport in the form of fluoride compounds, however, is quite easy. The nature of the solvent cation has a considerable influence on the intensity and (more important still) the direction of mass transfer; this indicates that the elements is question are transported in the form of fluoro-complexes of titanium, zirconium, and hafnium with alkali metals and ammonia, The recrystallization of TiO_2, ZrO_2, and HfO_2 may in this case be schematically represented as the formation of a corresponding compound of the type $X_x M_y F_z$ (X = K, Na, NH_4; M = Ti, Zr, HF) and the subsequent pyrolysis of this compound in the growth zone to form* MO_2.

The coefficients x, y, z in the formulas of the intermediate compounds are unknown, since titanium, zirconium, and hafnium may combine with fluorine in a variety of ways: MF^{3+}, MF_6^{2-}, MF_5^{1-} and others [10]. At high temperatures the MF_5^{1-} ions are more

---

*The transport of Ti, Zr, and Hf may also take place in the form of oxyfluoride compounds. This has no serious effect on the course of the discussions, and we shall in future always assume that transport occurs in the form of fluoro-complexes.

TABLE 3

Original reagents	No. of experiments	Temperature of experiments,°C	Solution	pH after experiment	Final products*
$(NH_4)_2 TiF_6$	3	600	H_2O	~4	TiO_2 (0.053—0.077 g), original reagent
$(NH_4)_2 TiF_6$	3	600	$H_2O + NH_4OH$	~7	TiO_2 (0.050—0.069 g), original reagent
$(NH_4)_2 TiF_6$	3	500	H_2O	~4	TiO_2 (0.03—0.039 g), original reagent
$(NH_4)_2 TiF_6$	3	500	$H_2O + NH_4OH$	~7	TiO_2 (0.019—0.031 g), original reagent
$(NH_4)_3 ZrF_7$	3	600	H_2O	~4	ZrO_2 (0.035—0.06 g), original reagent
$(NH_4)_3 ZrF_7$	2	600	$H_2O + NH_4OH$	~7	ZrO_2 (0.03—0.04 g), original reagent
$(NH_4)_3 ZrF_7$	3	500	H_2O	~4	ZrO_2 (0.01—0.02 g), original reagent
$(NH_4)_3 ZrF_7$	3	500	$H_2O + NH_4OH$	~7	ZrO_2 (0.01 g), original reagent
K_2ZrF_6	3	600	H_2O	~5	Traces of ZrO_2 original reagent
K_2ZrF_6	3	600	$H_2O + KOH$	~13	Complete hydrolysis
K_2ZrF_6	3	500	H_2O	~5	Traces of ZrO_2 original reagent
K_2ZrF_6	3	500	$H_2O + KOH$	~13	Complete hydrolysis

*Maximum and minimum weight of the oxides found in the experiments under identical conditions.

stable,* and this form may well represent the closest approach to the form of transport of titanium, zirconium, and hafnium under hydrothermal conditions.

According to the foregoing scheme of TiO_2, ZrO_2, and HfO_2 recrystallization, the direction of transport of the oxides is determined by the temperature dependence of the hydrolysis constant of the fluoro-complexes. In the case of ammonium fluorotitanates, fluorozirconates, and fluorohafnates, in particular, it would appear that the hydrolysis constant increases with rising temperature under the conditions envisaged, and this is why TiO_2, SrO_2, and HfO_2 move in the direction of the "hot" zone of the autoclave. The hydrolysis constant of potassium and sodium fluorozirconates,

---

*In the case of ammonium fluorozirconates the following reaction occurs at high temperatures:

$$(NH_4)_3ZrF_7 \xrightarrow{296° C} (NH_4)_2ZrF_6 \xrightarrow{357° C} NH_4ZrF_5.$$

fluorotitanates, and fluorohafnates evidently falls with increasing temperature.

Accurate data relating to the hydrolysis constants of the fluoro-titanates, fluorozirconates, and fluorohafnates at high temperatures and pressures might well serve as a confirmation of the foregoing assumptions as to the transport forms and mechanisms of titanium, zirconium, and hafnium. However, no direct measurements of the hydrolysis constants of these compounds have ever yet been carried out under the conditions of present interest. In order to make an indirect estimate of the rate of hydrolysis, we therefore studied the stability of a number of fluorotitanates and fluorozirconates under hydrothermal conditions. To this end we employed the following method. The compound in question is first held for a long period in an autoclave in sealed ampoules containing a solution with a prespecified pH value. Then the reaction products are analyzed. The presence of oxygen compounds among these (in the form of titanium, zirconium, or hafnium oxides or hydroxides) indicates the hydrolysis of the original fluoro-complexes, and the amount of oxides so formed qualitatively characterizes the intensity of the process. The period of the experiment is made long enough to ensure equilibrium in the ampoules (in our own case four days).

As original reagents we used $(NH_4)_2TiF_6$, $(NH_4)_3ZrF_7$ and K_2ZrF_6. The whole of the original charge was 0.5 g. The principal conditions and results of the experiments are presented in Table 3.

We see from Table 3 that ammonium fluorozirconates are less stable under hydrothermal conditions than potassium fluorozirconates. For the latter compounds, hydrolysis only occurs at high pH values ($\sim$ 13), whereas ammonium fluorozirconates and fluorotitanates hydrolyze even in pure water. The amount of precipitating TiO_2 and ZrO_2 is smaller at 500 than at 600°C, which may be regarded as a qualitative confirmation of our conclusion regarding the increasing hydrolysis constant of ammonium fluoro-titanates and fluorozirconates with increasing temperature. No specific conclusions may be drawn from the experiments as to the temperature dependence of the hydrolysis constant of potas-sium fluorozirconate.

Conclusion

The hydrothermal recrystallization of titanium, zirconium, and hafnium oxides takes place with the greatest intensity in fluoride solutions of potassium, sodium, and ammonium. This distinguishes TiO_2, ZrO_2, and HfO_2 from a large group of oxides (SiO_2, Al_2O_3, ZnO, Bi_2O_3, GeO_2, HgO, etc.) which readily undergo hydrothermal recrystallization in alkali solutions. The transport of titanium, zirconium, and hafnium in fluoride solutions takes place in the form of fluoro-complexes, while the direction of transport is determined by the temperature dependence of the hydrolysis constant of these compounds. The stability of the latter under hydrothermal conditions, in turn, depends on the cation of the solvent employed.

The recrystallization of TiO_2, ZrO_2, and HfO_2 is "restricted" to relatively high temperatures and a small number of solvents; in the case of ZrO_2 and HfO_2, solutions of NH_4F are practically the only serviceable ones. These restrictions also operate in the crystallization of complex systems containing titanium and zirconium as constituents. The existence of other components in the complex systems will introduce further correcting factors; however, as we shall show in a subsequent article [11], the recrystallization of certain titanates and zirconates takes place under conditions similar to those of the recrystallization of "pure" titanium and zirconium oxides.

Literature Cited

1. Handbook of Chemistry [in Russian], Vol. 2, Izd. Khimiya (1964).
2. H. W. Newkirk and D. K. Smith, Amer. Mineral., 50:44 (1955).
3. C. Doelter, Handbuch der Mineral-Chemie, Vol. 4, Pt. 1, T. Steinkopff, Vienna, (1926).
4. A. B. Chase and J. A. Osmer, Amer. Mineral., 51:1808 (1966).
5. P. Mergault and G. Blanche, Compt. Rend., 238:914 (1954).
6. T. Sugai, S. Hesegawa, and G. Ohara, J. Appl. Phys. Japan, 6:901 (1967).
7. D. D. Dena, E. S. Dena, Ch. Pelach, G. Berman, and K. Frondel', Sistema Mineralogii, 1(2):94 (1951).
8. I. N. Anikin, I. A. Naumova, and G. V. Rumyantseva, Kristallografiya, 10:231 (1965).
9. M. L. Harvill and R. Roy, J. Phys. Chem. Solids, Suppl. 1, p. 563 (1967).
10. G. Ryss, Chemistry of Fluorine and Its Inorganic Compounds [in Russian], Goshimizdat (1956).
11. V. A. Kuznetsov, this volume, p. 81.

Synthesis of Single Crystals of Ternary Chalcohalides

V. I. Popolitov and B. N. Litvin

The chalcohalides of antimony and bismuth had already been synthesized before the end of the last century [1, 2]. However, only after the structure of these compounds had been studied by x-ray diffraction did the study of their optical, dielectric, and photo-electric properties really begin [3-6]. Of all the chalcogenides of antimony and bismuth, that which has atracted the greatest attention from research workers has been antimony sulfoiodide, which possesses both ferroelectric and photosemiconducting properties. Methods of producing chalcogenides as well as the external appearance of the compounds so formed, their physical and chemical properties, are described in [7]. Chalcohalides may be obtained (1) by the limited action of dry hydrogen sulfide on a heated (but not melted) halide (2) by melting the corresponding sulfides and halides together, and (3) by the interaction of sulfides with halides. It should be noted that the composition of the reacting mixtures is usually chosen empirically, without any real physicochemical basis for the conditions of synthesis. In order to establish a reasonable basis for this, it is essential to study the phase diagrams of the chalcogenide–halide systems M_2S_3–X_3, where M = Sb or Bi, X = Cl, Br, or I, as well as the chalogenide-chalcogen X_3–S(Se), the metal–chalcogen–halogen, and the chalcogenide-halogen systems.

The best-known papers which deal with this subject are those of Japanese [8] and Soviet [9] scientists. Part of the phase diagram of the ternary Sb–S–I system was studied in [8]. The authors in question investigated the $Sb_2S_3-Sb_2I$ system along the cooling curves by thermal analysis. In order to determine the conditions governing the formation of the sulfoiodide by the interaction of antimony sulfide with iodine or the iodide, the phase diagrams of the binary systems $Sb_2S_3-SbI_3$ and $Sb_2S_3-I_2$ were studied in [9] by the thermal-analysis method. The phase diagrams of the $Sb_2S_3-SbI_3$ and $Sb_2S_3-I_2$ systems were plotted from the thermal analyses.

V. A. Lyakhovitskaya and L. M. Belyaev obtained antimony sulfoiodide from the gas phase [10]. Crystallization from the gas phase took place in a closed vessel at 360°C (evaporation zone) and 320°C (crystallization zone). The original charge consisted of antimony sulfoiodide and a certain amount of antimony iodide. In this article the authors also describe the mechanism underlying the formation of antimony sulfoiodide from the gas phase.

Thus methods of producing antimony and bismuth chalcohalides either by direct melting or by way of transport reactions had been developed well before our own series of investigations.

The principal experimental difficulty arising when studying the conditions of synthesis of the ternary chalochalides $A^VB^{VI}C^{VII}$ under hydrothermal conditions lay in creating conditions such as would eliminate the possibility of oxidation. In other words, it was essential that the original components should have the valence states $A^{+3}B^{-2}C^{-1}$ in the solution. First Eh–Ph diagrams of the individual A^V-H_2O, $B^{VI}-H_2O$, $C^{VII}-H_2O$ systems were constructed, and by superposing these (Fig. 1) the region (shaded) in which all three ions were in the required valence states was obtained. Analogous regions were plotted on the Eh–pH diagrams for the systems $As-S-I-H_2O$ (Fig. 2) and $Bi-S-I-H_2O$ (Fig. 3). The substitutions $S \rightarrow Se \rightarrow Te$ and $I \rightarrow Br \rightarrow F$ introduced very little distortion into the region of simultaneous existence of the ions A^{+3}, B^{-2}, and C^{-1}. We see from an inspection of the diagrams that the existence of all three ions in the required valence state is determined by the strongly reducing properties of the medium. The appearance of metallic As, Sb, or Bi during the synthesis will indicate that the lower limit of the redox potential has been passed (curve AB). The upper limit is the line of stability of the sulfur ions S^{2-} (curve CD). However, a specific value of the redox potential is merely a necessary, not a sufficient condition for the synthesis of $A^VB^{VI}C^{VII}$. Apart from the ternary chalcohalides, compounds of the $A^VB_3^{VII}$, $A_2^VB_3^{VI}$, and other types may be formed in

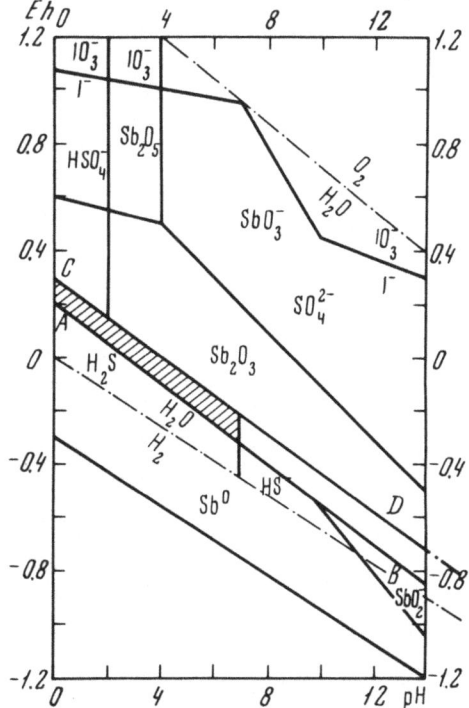

Fig. 1. Eh–pH diagrams of the Sb–S–I–H₂O system.

the region under consideration. The probability of the formation of any of these compounds will be determined by the concentration of the original components, as well as the pH of the medium. Let us consider some possible alternatives. Depending on the pH of the medium, sulfur forms the following complexes:

$$[H_2S] \rightleftarrows [HS^-] \rightleftarrows [S^{2-}],$$
$$\text{pH}<7 \quad \text{pH}=7-10 \quad \text{pH}>10$$

In the presence of sulfur, in an alkaline medium, antimony is characterized by a complex of the $[SbS_2^-]$ type, and in an acid medium, in the presence of a halogen such as chlorine, by one such as $-[SbCl_4^-]$. For all other combinations of the ABC type a close analogy may be drawn. Interaction between the developing complexes should lead to the following reaction (taking SbSI as an example):

$$\text{pH}<6 \quad \begin{cases} Sb + 4HI \rightarrow [SbI_4^-] + 2H_2 \uparrow, \\ 2\,[SbI_4^-] + 4\,[H_2S] \rightarrow [Sb_2I_2\,(H_2S)_4^{4-}] + 6I, \\ [Sb_2I_2\,(H_2S)_4^{4-}] \rightarrow 2SbSI + 2H_2S + 2H_2 \uparrow. \end{cases}$$

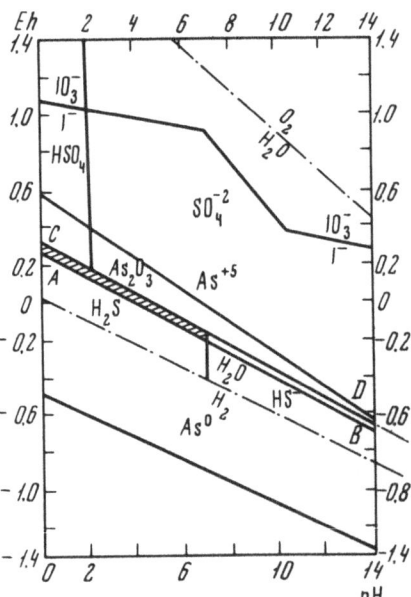

Fig. 2. Eh–pH diagrams of the As–S–I–
H₂O system.

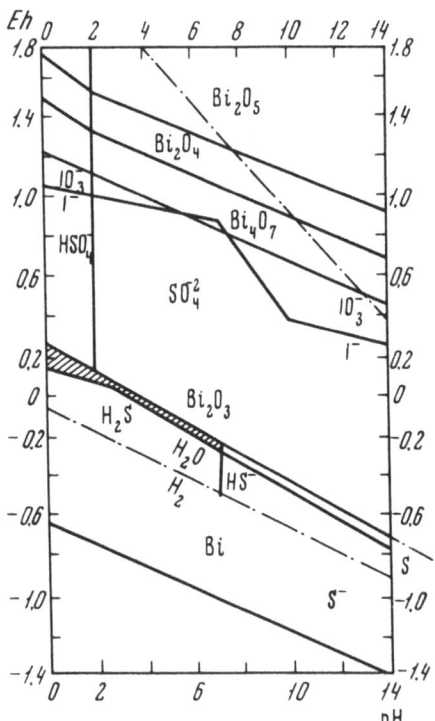

Fig. 3. Eh–pH diagrams of the
Bi–S–I–H₂O system.

$$pH = 7-10 \begin{cases} Sb + 4HI \rightarrow [SbI_4^-] + 2H_2 \uparrow, \\ [SbI_4^-] + 2[HS^-] \rightarrow [SbS_2^-] + 2I + 2HI, \\ Sb + 2[HS^-] \rightarrow [SbS_2^-] + H_2 \uparrow, \\ 2[SbS_2^-] + 4[HS^-] \rightarrow 2Sb_2S_3 + 2H_2S \uparrow. \end{cases}$$

$$pH > 10 \begin{cases} Sb + 2[HS^-] \rightarrow [SbS_2^-] + H_2 \uparrow, \\ 2[SbS_2^-] + 4[HS^-] \rightarrow 2Sb_2S_3 + 2H_2S \uparrow, \\ 2[SbS_2] + [S^{2-}] \rightarrow Sb_2S_3 + 2S. \end{cases}$$

Hence SbSI will chiefly be formed in solutions with pH < 7. The range of formation of antimony sulfoiodide can only be established by careful experiment. The formation of SbSI in solutions with low pH values may be associated with the stability of a hypothetical complex of the $[Sb_2I_2 (H_2S)_4^{4-}]$ type, which may be formed as a result of an interaction between the two genuinely existing complexes $[SbI_4^-]$ and $[H_2S]$. In neutral and alkaline solutions the sulfo-complex $[SbS_2^-]$ is well known to be stable; on interaction with other sulfide complexes it necessarily leads to the formation of Sb_2S_3.

The complex $[Sb_2I_2 (H_2S)_4^{4-}]$ may be represented in the form

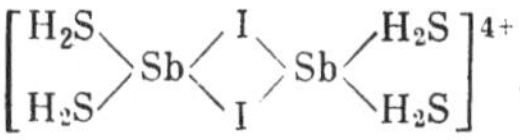

$$\left[\begin{array}{c} H_2S \\ H_2S \end{array} \!\!\!\! Sb \!\! \left\langle \begin{array}{c} I \\ I \end{array} \right\rangle \!\! Sb \!\!\!\! \begin{array}{c} H_2S \\ H_2S \end{array} \right]^{4+},$$

and then the formation of SbSI should take place as a continuous interaction between such complexes (Fig. 4).

On the basis of the scheme presented in Fig. 4, the conditions for congruent dissolution and recrystallization may be determined from the equation

$$2SbSI + 2[H_2S] + 2H_2 \rightleftarrows [Sb_2I_2 (H_2S)_4^{4-}].$$

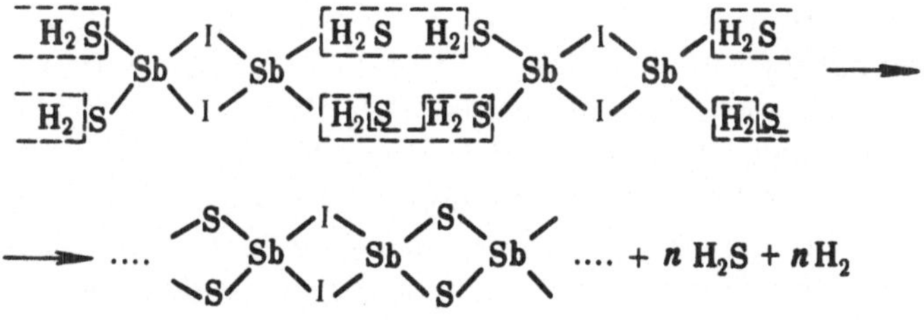

Fig. 4. Scheme of interaction between sulfoiodide complexes.

This type of derivation should be valid for the whole $A^V B^{VI} C^{VII}$ group:

$$2 A^V B^{VI} C^{VII} + 2 [H_2 B^{VI}] + 2H_2 \rightleftarrows [A_2^V C_2^{VII} (H_2 B^{VI})_4^{4-}].$$

This reaction subsequently served as a basis for developing a method of synthesizing and crystallizing the whole $A^V B^{VI} C^{VII}$ "family."

The investigation was carried out in autoclaves (made of 1Kh18N10T steel) using Teflon and titanium linings. As original charge we used the following chemical reagents: antimony and bismuth halides, metallic arsenic, antimony, and bismuth, elemental sulfur, selenium, and iodine, and sodium polysulfide; we also used the natural minerals $Sb_2 S_3$ and $As_2 S_3$. The solvents were aqueous solutions of hydrohalic acids: HF, HCl, HBr, and HI. The pH of the media was varied by varying the ratio between the corresponding hydrohalic acid and the sodium sulfide. All the solid components of the charge except for the sodium sulfide were introduced into the space inside the lining as a mechanical mixture. The sodium sulfide was placed in a special small cylinder in the upper part of the lining and entered into reaction with the hydrohalic acid during the synthesis. Figure 5 illustrates a cross section of the autoclave with a floating lining ready for the experiments.

We studied crystallization in the Sb–S–I–H$_2$O system. The yield of SbSI crystals was taken as the principal index of the hydrothermal reactions. The results of the experiments are presented in Table 1.

In the first series of experiments (Table 1) we used elemental antimony, sulfur, and iodine or an iodine solution as charge. The maximum yield of SbSI occurred for an Sb : S : I ratio of 2 : 1.5 : 2 (experiment 6). On increasing the Sb/(S + I) ratio, fine isometric crystals of native antimony 1–2 mm in size appear. An excess of sulfur (experiment 4) increases the partial pressure of hydrogen sulfide and promotes the formation of crystallites of elemental sulfur (about 0.01 mm in size). The main characteristics of this series are the formation of reddish SbI$_3$ single crystals (5–15%) and the poor solubility of the antimony. With increasing iodine concentration (falling pH) the solubility of the antimony increases; however, this does not lead to any increase in the yield of SbSI, but tends to promote the formation of SbI$_3$ crystals. The stable formation of antimony triiodide may be explained by the appearance

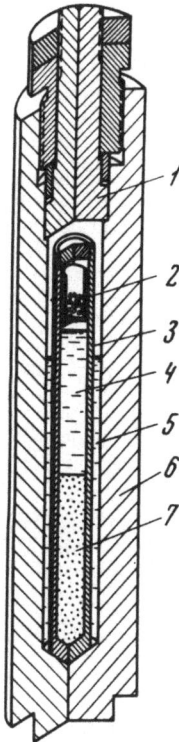

Fig. 5. Arrangement of the autoclave in the working state. 1) Obturator; 2) small cylinder containing Na_2S; 3) floating lining; 4) aqueous solution of iodine; 5) water; 6) body of autoclave; 7) charge.

of the complex $[SbI_4^-]$ under the conditions in question, as a result of which some of the antimony is strongly bound into a compound. If we now introduce another halogen ion such as chlorine into the solution in addition to iodine, this should sharply reduce the yield of of SbSI and also influence the yield of SbI_3. This behavior of SbSI and SbI_3 is associated with the formation of a strong $[SbCl_4^-]$ complex.

In experiments 11-14 chlorine was introduced into the charge in the form of $SbCl_3$; the SbSI yield diminished and crystallization of the SbI_3 rarely occurred. On the other hand, preliminary strong interaction between antimony and sulfur should increase the yield of SbSI, since this eases the formation of the complex $[Sb_2I_2(H_2S)_4^{4-}]$. In experiments 7-9 we used Sb_2S_3 in the charge; the yield of SbSI remained at the same level as in the experiments with elemental antimony and sulfur, while the yield of SbI_3 diminished. Thus

TABLE 1

Experiment	Composition of charge	Weight relationships of charge components	Results of crystallization, %		
1	Sb, S, I	1 : 1 : 1	SbSI ~15,	SbI₃ 10,	some of the incompletely-reacting Sb
2		1 : 1 : 1	17,	8,	some of the incompletely-reacting Sb
3		2 : 1 : 1	18,	10,	single crystals Sb
4		1 : 2 : 1	21,	SbI₃ 5,	sulfur, H₂S, single crystals Sb
5		1 : 1 : 2	25,	11,	solid iodine, Sb
6		2 : 1,5 : 2	31,	15,	sulfur, H₂S, single crystals Sb
7	Sb₂S₃, I	1 : 1	18,	7,	single crystals Sb
8		1 : 2	24,	12,	solid iodine
9		2 : 1	19,	9,	sulfur, single crystals Sb
10	Sb₂S₃, S, I	1 : 0,2 : 1	20,	10,	sulfur
11	SbCl₃, S, I	1 : 1 : 1	4—5,		sulfur, single crystals Sb
12		1 : 2 : 1	7—8,		sulfur, single crystals Sb
13		1 : 1 : 2	8—10,	SbI₃	2—3, solid iodine
14		2 : 1 : 1	3—5,		Sb single crystals, traces of sulfur

Sb_2S_3 has no favorable effect on the yield of SbSI, at any rate in solutions with pH = 2-3.

In the experiments with a high chlorine concentration and a low pH value (1.5), the amount of SbSI diminishes sharply (Table 1, experiments 1-14, and Table 2, experiments 9-10). The principal phases become SbI_3, Sb_2S_3, sulfur, and crystals of metallic antimony.

In the second series of experiments we used solutions with a high pH value (4-7), introducing Na_2S into the solution for this purpose. The results of the experiments are presented in Table 3. Increasing the pH of the solvent greatly increased the yield of SbSI (70-75%). An excess of antimony in the charge leads to the crystallization of Sb_2S_3. A characteristic feature of all the experiments with Na_2S is the appearance of sulfur crystals of a yellowish color (experiments 1-12).

Finally in the third series of experiments we studied the effect of the partial pressure of hydrogen sulfide on the crystalliza-

TABLE 2

Ex-peri-ment	Composition of charge	Weight rela-tionships of charge components	Concen-tration of HCl, %	Results
1	Sb, S, I	1 : 1 : 1	3	SbSI $\sim$ 10—12%,sulfur, iodine, antimony
2		1 : 1 : 1	20	SbSI$\sim$3—4%, sulfur, iodine, antimony
3		1 : 1 : 1	35	SbI$_3$ $\sim$ 15—16%, sulfur, iodine, Sb$_2$S$_3$ single crystals
4		1 : 2 : 1	10	SbSI$\sim$9—10%, sulfur, iodine, antimony
5		2 : 1 : 1	35	SbI$_3\sim$ iodine, single crystals of Sb$_2$S$_3$
6		1 : 1 : 2	20	SbSI$\sim$5—6%,sulfur, iodine, Sb$_2$S$_3$ single crystals
7	Sb$_2$S$_3$, I	1 : 1	9	SbSI$\sim$9—10%, iodine, traces of sulfur
8		1 : 1	35	SbI$_3\sim$15—21%, iodine, sulfur, Sb$_2$O$_3$
9	SbCl$_3$, S, I	1 : 1 : 1	5	SbSI $\sim$7—9%, iodine, antimony single crystals
10		1 : 1 : 1	35	SbI$_3$ $\sim$ 15—16%, sulfur, single crystals of Sb, Sb$_2$S$_3$

TABLE 3

Ex-peri-ment	Composition of charge	Weight rela-tionships of the charge components	Amount of Na$_2$S, g	Results
1	Sb, S, I	1 : 1 : 1	2	SbSI $\sim$37—40%, sulfur, single crystals of Sb
2		1 : 1 : 1	10	SbSI $\sim$63—66%, sulfur, Sb$_2$S$_3$
3		2 : 1 : 1	8	SbSI $\sim$53—55%, sulfur, Sb$_2$S$_3$, single crystals of Sb
4		1 : 2 : 1	8	SbSI $\sim$56—59%, sulfur, Sb$_2$S$_3$
5		1 : 1 : 2	8	SbSI $\sim$70—75%, SbI$_3$, iodine
6	Sb$_2$S$_3$, I	1 : 1	8	SbSI $\sim$ 60%, single crystals of Sb
7		1 : 2	8	SbSI $\sim$70—72%, SbI$_3$, iodine
8		2 : 1	8	SbSI $\sim$65—67%, single crystals of Sb, Sb$_2$S$_3$
9	SbCl$_3$, S, I	1 : 1 : 1	8	SbSI $\sim$30—32%, sulfur, single crystals of Sb
10		2 : 1 : 1	8	SbSI $\sim$43—45%, Sb$_2$S$_3$. single crystals of Sb
11		1 : 2 : 1	8	SbSI $\sim$46—48%, Sb$_2$S$_3$, sulfur
12		1 : 1 : 2	8	SbSI$\sim$63—65%, Sb$_2$S$_3$, iodine

TABLE 4

Experiment	Weight relationships of charge components Sb : S : I	Amount of Na₂S, g	Concentration of HCl, %	Results
1	2 : 1 : 1	5	35	SbSI $\sim$ 17—19%, Sb₂S₃, single crystals of Sb, sulfur
2	2 : 1 : 1	10	7	SbSI $\sim$ 90—92%, single crystals of Sb
3	2 : 1 : 1	10	35	SbSI $\sim$ 26—28%, Sb₂S₃, single crystals of Sb
4	1 : 1 : 1	5	7	SbSI $\sim$ 100%, H₂S
5	1 : 2 : 1	5	35	SbSI $\sim$ 18—20%, sulfur, Sb₂S₃, H₂S
6	1 : 2 : 1	10	7	SbSI $\sim$ 95%, sulfur, H₂S
7	1 : 2 : 1	10	35	SbSI $\sim$ 26—28%, Sb₂S₃, H₂S
8	1 : 1 : 2	5	7	SbSI $\sim$ 72—75%, solid iodine
9	1 : 1 : 2	5	35	SbSI $\sim$ 19—20%, iodine, Sb₂S₃, single crystals of Sb, SbI₃, sulfur
10	1 : 1 : 2	10	7	SbSI $\sim$ 80—82%, Sb₂S₃, iodine, SbI₃
11	1 : 1 : 2	10	35	SbSI $\sim$ 27—29%, Sb₂S₃, iodine, single crystals of SbI₃, Sb, sulfur

tion of SbSI. The formation of hydrogen sulfide in the reactor took place in accordance with the reaction $Na_2S + 2HCl = 2NaCl + H_2S\uparrow$. The construction of the reactor did not allow the partial pressure of the H_2S to be measured directly; hence the quantitative characteristic was the amount of hydrogen sulfide evolved in accordance with the foregoing reaction. The results of the experiments are presented in Table 4. The yield of SbSI becomes almost 100% in aqueous solution with an average pH of 5-6 and a low partial pressure of hydrogen sulfide. Clearly the next most important step would be a detailed study of the stability and recrystallization of antimony sulfoiodide in the complex system H_2S—SbSI—H_2O at various pH values, these latter being specified independently. A special reactor designed to initiate such an investigation is now being developed.

The results obtained in connection with the conditions of SbSI synthesis enabled us to synthesize the isomorphic varieties of the whole family of chalcohalides. The isomorphic substitutions were based on the following schemes: Sb $\rightarrow$ As $\rightarrow$ Bi; S $\rightarrow$ Se; I $\rightarrow$ Br $\rightarrow$ Cl–F. Assuming that these substitutions make little

TABLE 5

Experiment	Composition of charge	Weight relationships of charge components	Additional charge components, g		Solvent, wt.%	Results
			Na₂S	Na₂Se		
1	SbCl₃, S	1:1	10	—	HCl, 6	SbSCl, Sb, sulfur
2	Sb, S	1:1	10	—	HBr, 6	SbSBr, Sb, sulfur
3	Sb, S	1:1	10	—	HF, 7	SbSF, Sb, sulfur
4	Sb, Se, I	1:1:1	—	10	HCl, 5	SbSeF, Sb
5	Sb, Se	1:1	—	10	HBr, 5	SbSeBr, Sb
6	Sb, Se	1:1	—	10	HF, 5	SbSeF, Sb
7	Bi, S, I	1:1:1	10	—	HCl, 10	BiSI
8	Bi, S, I	1:1:1	2	—	HCl, 15	~10% BiSI, BiI₃, iodine, sulfur
9	Bi, S, I	1:1:1	5	—	HCl, 25	~8–9% BiSI, BiI₃, iodine, sulfur
10	Bi, S, I	1:1:1	7	—	ICl, 12	31–32% BiSI, BiI₃, iodine, sulfur
11	Bi, S	1:1	10	—	HBr, 6	BiSBr, sulfur, bismuth
12	Bi, S	1:1	10	—	HF, 5	BiSF, sulfur, bismuth
13	BiCl₃, S	1:1	10	—	HCl, 5	BiSCl, sulfur, bismuth
14	As₂S₃, I	1:1	10	—	HCl, 7	AsSI, arsenic
15	As₂S₃, I	2:1	10	—	HCl, 7	AsSI, arsenic, sulfur

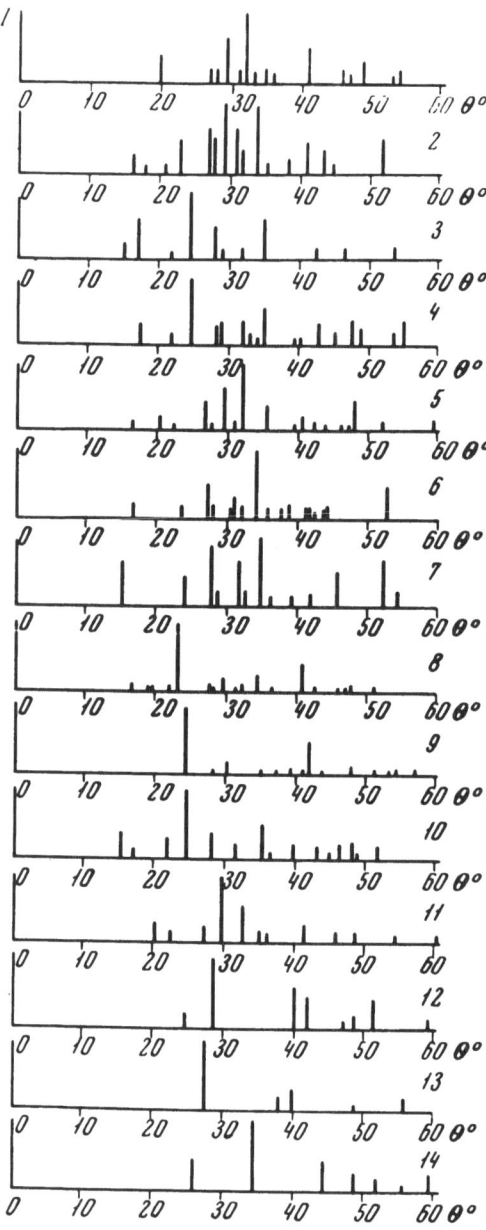

Fig. 6. Line diagrams of the phases obtained: 1) SbSI; 2) SbSBr; 3) SbSCl; 4) SbSF; 5) SbSeI; 6) SbSeBr; 7) SbSeCl; 8) BiSI; 9) BiSBr; 10) BiSCl; 11) AsSi; 12) Sb; 13) Bi; 14) As.

difference to be boundaries of the hypothetical range of stability of the ternary chalcohalides, the synthesis of the isomorphic compounds reduces to a simple change of charge components. As in the case of the SbSI, the yield of crystals in this case depended on the pH of the medium and the partial pressure of H_2S or H_2Se. Naturally the synthesis of each compound required a certain change in the experimental conditions. For example, in order to obtain SbSeI it was sufficient simply to replace Na_2S by Na_2Se in the charge (experiments 4-6 in Table 5). However, the replacement of iodine by bromine, chlorine, and fluorine involved a special choice of the concentrations of the corresponding solutions of HBr, HCl, and HF.

The method of producing arsenic and bismuth sulfohalides differed very little from that of producing the corresponding antimony sulfohalides. Above 200°C the arsenic varieties are characterized by the "igneous" melting of the charge, as a result of which a reddish-brown viscous liquid is formed. In experiments at about 300°C the liquid turns into a vitreous mass on cooling, and this is amorphous with respect to x rays.

All the chalcohalides obtained were studied by x-ray diffraction, and the corresponding line diagrams are presented in Fig. 6. We shall now give a short description of the phases so synthesized.

SbSI. The crystals are dark red and reddish brown in color, with metallic iridescence. The principal shape is acicular, the needles being 30-40 mm long and 0.1-1.0 mm thick. In individual experiments the crystals reached 130-150 mm. For a large tem-

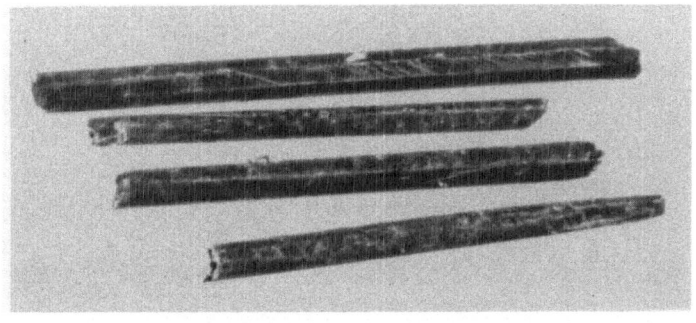

Fig. 7. Single crystals of SbSI (× 8).

perature drop $\Delta T > 20°C$ short, columnar crystals are formed (Fig. 7). In solutions with pH = 5-6, SbSI crystallizes in the form of individual, well-formed varieties of maximum length. In more acid solutions, shorter, thinner needles are obtained, these frequently being combined into aggregates forming spherulites and "pencils" (Fig. 8).

A s S I. Arsenic sulfoiodide crystallizes in the form of a mass of short (up to 5 mm) fine (0.05 mm) needles of a reddish color. Changes in the pH of the solution affect AsSI less.

B i S I. The crystals are acicular, opaque, and of a dark color (Fig. 9).

The replacement of S by Se leads to the formation of short-columnar, opaque crystals. The Cl and F compounds crystallize in the form of thin needles with a metallic luster. The sulfo-bromides form short-columnar crystals (Fig. 10) of dark brown color.

In cases in which the redox potential is sufficiently low, crystals of the semimetals aresenic, bismuth, and antimony are formed. The yield of these was usually quite low (5-10%), and we made no special attempts at determining the conditions required for 100% crystallization. Arsenic crystals of rhombohedral habit

Fig. 8. Spherulites of SbSI ($\times$ 8).

Fig. 9. Crystals of BiSI (×2).

and dark gray color are formed in conjunction with AsSI, the luster being metallic and the dimensions about 1-2 mm.

Antimony crystallizes in the form of isometric, lamellar varieties of light gray color; these are brittle, and up to 3-4 mm in size. Bismuth crystals are similar to those of antimony.

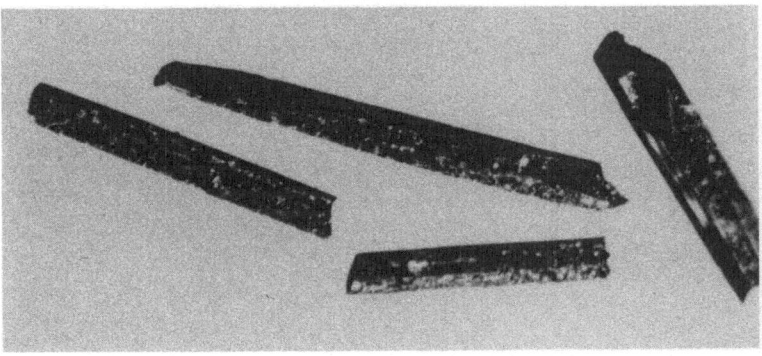

Fig. 10. Crystals of SbSeBr (×4).

Conclusions

1. By considering the Eh–pH diagrams relating to the stability of the ions in systems of the A–S–I–H_2O type (where A = As, Sb, or Bi) we have demonstrated the possibility of synthesizing ASI single crystals in aqueous solutions.

2. By analyzing the theoretical requirements, we have established the conditions underlying the crystallization of SbSI, in so far as these relate to the temperature, the ratio of the original charge components, the partial pressure of the H_2S, the pH of the medium, and complexing processes. The intensity of the crystallization of SbSI depends on which of the foregoing factors happens to be dominant.

3. The results obtained in relation to the conditions of crystallization of SbSI have enabled us to synthesize the isomorphic analogs as well. The isomorphic analogs are obtained by simply changing the composition of the charge and the solvents.

Literature Cited

1. L. Ouvrard, Compt. Rend., 1:107 (1893).
2. R. Schneider, Pogg. Ann., 109:610 (1860).
3. E. Fatuzzo and G. Harbere, Phys. Rev., 127:2036 (1962).
4. R. Kern, J. Phys. Chem. Solids, 23:249 (1962).
5. R. Armandt and A. Niggli, Naturwiss., 7:51 (1964).
6. R. Nitshe and W. Merz, J. Phys. Chem. Solids, 13:154 (1960).
7. M. V. Mokhoseev and S. M. Aleikina, in: Chalcogenides [in Russian], Izd. Naukova Dumka (1967).
8. T. Mori and H. J. Tamura, J. Phys. Soc. Japan, 19:1247 (1964).
9. M. V. Mokhoseev, S. M. Aleikina, and A. M. Gamol'skii, in: Chalcogenides [in Russian], Naukova Dumka (1967).
10. V. A. Lyakhovitskaya and L. M. Belyaev, Izv. Akad. Nauk SSSR, Neorgan. Mat., 10:1287 (1965).

Crystallization of Antimonite (Sb$_2$S$_3$)

V. I. Popolitov and B. N. Litvin

In this article we shall consider the crystallization of anti-monite, Sb$_2$S$_3$, under hydrothermal conditions in aqueous solutions of chlorides and Na$_2$S. We studied the crystallization of Sb$_2$S$_3$ in ordinary autoclaves with floating titanium linings having knife-edge bearings. As solvent we used aqueous solutions of NH$_4$Cl, NH$_4$Br, and Na$_2$S in concentrations between 2 and 12 wt.%. The solvent was poured into the lining with an autoclave occupation factor of 0.7. The experiments were carried out at 300 and 400°C; the temperature drop was kept constant at 20°C. The duration of the experiments was times so as to leave some unspent mixture in the lining after the experiments had ended. The residue of in-completely-reacted material indicated that the process had been interrupted while material was still being supplied to the growi crystal. The rate of crystal growth could therefore be determined by dividing the thickness of the layer by the time of the experiment, which was kept constant at six days. The rate of growth of the (001) face was determined by dividing the length of the crystal by the time of the experiment. The rate of growth in a direction perpendicular to the needle axis, or in the [001] direction, was determined by dividing the width of the face by the time of the experiment. In each experiment we selected 100 crystals and measured them under binoculars. The results of the measurements were analyzed statistically, using the equations of [1] (Table 1).

Figure 1 illustrates the rate of growth of the (001) face as a function of solvent concentration. This may be expressed by a simple linear equation:

$$v = kC + B_0,$$

73

TABLE 1

Experi- ment	Tempera- ture, °C	Solvent con- centration, %	Rate of growth of the (001) face, mm/day	Rate of growth of the (011) face, 10^{-2} mm/day	Ratio of growth rates (001)/(011)
1	400	NH₄Br, 2	0.28	1.9	14
2	400	NH₄Br, 5	0.45	2.0	22.5
3	400	NH₄Br, 8	0.50	2.3	22.5
4	400	NH₄Br, 12	0.80	2.5	3.2
5	400	NH₄Cl, 2	0.41	2.1	19
6	400	NH₄Cl, 5	0.75	2.9	25
7	400	NH₄Cl, 8	1.03	4.2	24
8	400	Na₂S, 2	0.25	1.2	20
9	400	Na₂S, 5	0.33	1.5	22
10	400	NH₄Cl, 12	1.27	5.0	25
11	400	Na₂S, 8	0.41	1.8	22.5
12	400	Na₂S, 12	0.55	2.1	25
13	300	NH₄Cl, 2	0.20	1.0	20
14	300	NH₄Cl, 5	0.23	1.25	22.4
15	300	NH₄Cl, 8	0.40	1.5	26
16	300	NH₄Cl, 12	0.53	2.0	26

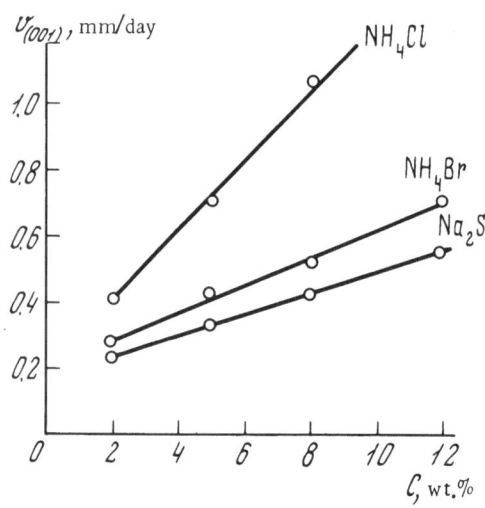

Fig. 1. Growth rate of the (001) face as a function
of the concentration of various solvents.

where v is the rate of growth in mm/day, k is the kinetic
coefficient, C is the solvent concentration in wt.%, and B_0 is a
constant, equal to 0.2 in the present case.

This empirical relationship characterizes the increase in
the rate of growth of the (001) face of Sb_2S_3 crystals on increasing

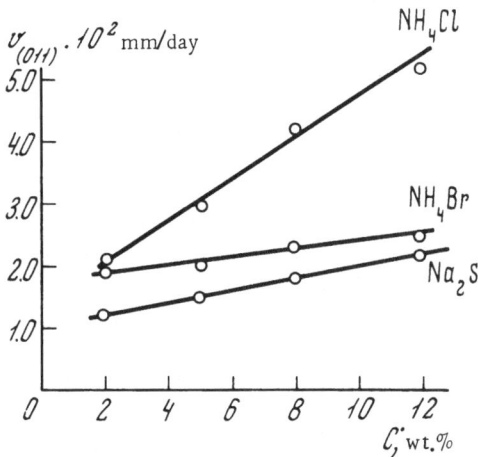

Fig. 2. Growth rate of the (011) face as a function
of the concentration of various solvents.

the concentration of the solvent by unity. The rate of growth of
the face increases linearly with increasing concentration of the
solvent, and more sharply in NH_4Cl than in NH_4Br and Na_2S solutions.

Figure 2 shows an analogous relationship for the growth rate
of the (011) face in NH_4Cl, NH_4Br, and Na_2S solutions. The relationship is described by the same equation and yields the following
order for the ratio of the kinetic coefficients: $k_{NH_4Cl} > k_{NH_4Br} >
k_{Na_2S}$. The growth rates of the (001) and (011) faces in NH_4Cl
solutions at 300 and 400°C are given in Fig. 3.

A rise in temperature increases not only the absolute
growth rate but also the value of the kinetic coefficient: $k_{400°C} >
k_{300°C}$.

The ratio of the absolute growth rates in NH_4Cl, NH_4Br, and
Na_2S solutions may be expressed by the inequality $v_{NH_4Cl} > v_{NH_4Br} >
v_{Na_2S}$, which is maintained over the whole concentration range for
both faces. The ratio of the growth rates for the two faces (001)/
(011), in turn, varies little with changing concentration; this reflects the constancy of the acicular habit of antimonite crystals
(Fig. 4).

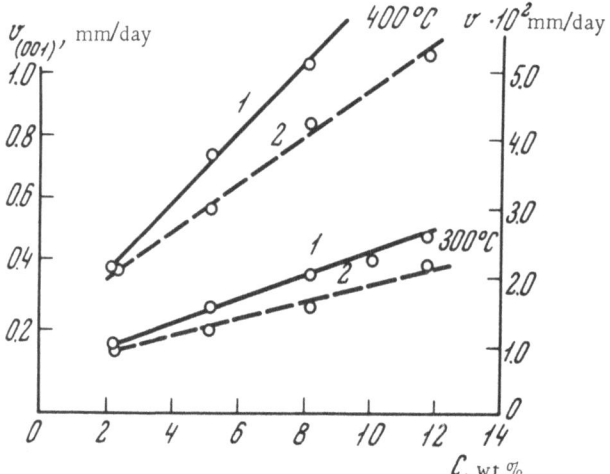

Fig. 3. Growth rates of the (001) face (1, left-hand scale)
and the (011) face (2, right-hand scale) at 300 and 400°C.

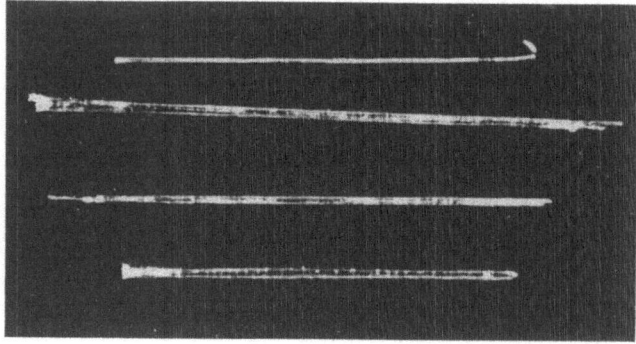

Fig. 4. Acicular crystals of antimonite ($\times$ 10).

In alkaline Na_2S solutions the original mixture will dissolve
in accordance with the following scheme [2]

$$Sb_2S_3 + 5HS' \rightleftarrows 2SbS_4^{5-} + 5H^{\cdot}$$

$$Sb_2S_3 + 2S'' \rightleftarrows SbS_2' + SbS_3^{3-}$$

$$\overline{2Sb_2S_3 + 5HS' + 2S'' \rightleftarrows SbS_2' + SbS_3^{3-} + 2SbS_4^{5-} + 5H^{\cdot}}$$

It follows from these equations that the formation of the complexes $[SbS_2]'$, $[SbS_3]^{3-}$ and $[SbS_4]^{5-}$ depends primarily on the concentration of $[HS]'$ and $[S]''$ ions in solution. For moderate pH values (8–10) the main complex will be $[HS]'$, while for pH > 10 $[S]''$ complexes will appear in the solution. Hence for pH 8–10 antimonite will dissolve with the predominant formation of complexes of the form

$$\left[\begin{array}{c} S \\ S \end{array} \!\! Sb \!\! \begin{array}{c} S \\ S \end{array} \right]^{5-},$$

and for pH > 10 complexes of the form

$$[S-Sb-S]^- \quad \text{and} \quad \left[S-Sb \begin{array}{c} S \\ S \end{array} \right]^{3-}.$$

In the crystallization zone these complexes will take part in the construction of the crystal lattice of Sb_2S_3. The structure of the latter constitutes a set of $[Sb_4S_6]\infty$ ribbons drawn out parallel to the c axis. In the same direction we find a preferential growth of the antimonite crystals, i.e., it would appear that the single crystals grow mainly by virtue of the attachment of particles to the [001] pinacoid surface, which lies perpendicular to the chains. The formation of antimonite in the crystallization zone may be pictured as follows

$$2SbS_4^{5-} + 5H^{\cdot} \rightarrow Sb_2S_3 + 5[HS]',$$

$$4 \left[\begin{array}{c} S \\ S \end{array} \!\! Sb \!\! \begin{array}{c} S \\ S \end{array} \right]^{5-} + 2H^{\cdot} \rightarrow \left[Sb \!\! \begin{array}{c} S \\ S \end{array} \!\! Sb \!\! \begin{array}{c} S \\ S \end{array} \!\! Sb \!\! \begin{array}{c} S \\ S \end{array} \!\! Sb \right]_{\infty} + 2HS',$$

or

$$SbS_2' + SbS_3^{3-} \rightarrow Sb_2S_3 + 2S'',$$

$$2\,[S-Sb-S] + 2 \left[S-Sb \begin{array}{c} S \\ S \end{array} \right]^{3-} \rightarrow \left[Sb \!\! \begin{array}{c} S \\ S \end{array} \!\! Sb \!\! \begin{array}{c} S \\ S \end{array} \!\! Sb \!\! \begin{array}{c} S \\ S \end{array} \right]_{\infty} + 2S''.$$

Since synthesis takes place in an aqueous medium at pH > 8, the mechanism underlying the interaction of the complexes with the growing surface should be associated with a dehydration reaction:

$$\cdots Sb \begin{array}{c} S-H_3O \\ S-H_3O \end{array} + \begin{array}{c} OH \\ OH \end{array} Sb \begin{array}{c} S\cdots \\ S\cdots \end{array} \rightarrow Sb \begin{array}{c} S \\ S \end{array} Sb \begin{array}{c} S \\ S \end{array} Sb \cdots + 4H_2O$$

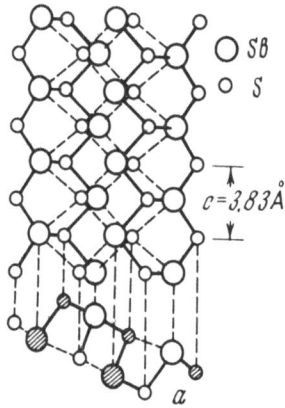

Fig. 5. Structure of the $[Sb_4S_6]_\infty$ strip of antimonite.

The antimony ions lie at different heights in the strip (Fig. 5). Hence at any given monent the (001) face accomodates hydrated antimony ions, which enter into chemical interaction with the complexes.

Let us now consider crystallization in an acid medium (aqueous solutions of NH_4Cl, and NH_4Br). Dissolution of the charge will take place in accordance with the following mechanism:

$$Sb_2S_3 + 8X' \rightleftarrows 2SbX_4' + 3S'' \text{ (where X = Cl, Br)}.$$

The inverse reaction, i.e., that of the formation of Sb_2S_3, involves the decomposition of the chloride complex:

$$\begin{bmatrix} X \\ X \end{bmatrix} Sb \begin{bmatrix} X \\ X \end{bmatrix}^- + 2S'' \rightarrow \begin{bmatrix} X \\ X \end{bmatrix} Sb \begin{bmatrix} X \\ X \end{bmatrix} + 2S'' + 2X' \rightarrow \begin{bmatrix} S \\ S \end{bmatrix} Sb \begin{bmatrix} S \\ S \end{bmatrix}^{5-} + 4X'.$$

The decomposition of the chloride complex, of course, takes place on the growing surface (topochemical reaction) and has the nature of a continuous interaction between the surface and the complex:

$$\cdots Sb \begin{matrix} S - H_3O & X \\ & - & \\ S - H_3O & X \end{matrix} Sb \begin{matrix} X \\ X \end{matrix} \rightarrow \cdots Sb \begin{matrix} S \\ S \end{matrix} Sb \begin{matrix} X \\ X \end{matrix} + 2S'' \rightarrow \cdots Sb \begin{matrix} S \\ S \end{matrix} Sb \begin{matrix} S - H_3O \\ S - H_3O \end{matrix} + \cdots$$

For a fairly high chloride concentration the topochemical reaction
will take place without dehydration:

Any disruption of the foregoing mechanism of the growth of the
(001) face of antimonite leads to the distortion or complete cessation
of the "chain" character of the growth process, and this causes
pyramid faces to appear.

Conclusions

　　1. Using mathematical statistics, we have calculated the
growth rate of the pinacoid [001] and prism [011] faces of Sb_2S_3
crystals in relation to the concentration of NH_4Cl, NH_4Br and Na_2S
at 400°C, and have shown that the growth rates of these faces may
be described by a simple linear equation of general form

$$v = kC + B_0.$$

　　2. The growth rates of the faces obtained from solutions of
NH_4Cl, NH_4Br and Na_2S may be expressed by the sequence $v_{NH_4Cl} >$
$v_{NH_4Br} > v_{Na_2S}$, this order being maintained over the whole of the
concentration range studied. The ratio of the growth rates of the
two faces (001)/(011) varies little with changing concentration in
solutions of NH_4Cl, NH_4Br and Na_2S; this reflects the constancy of
the acicular habit of the Sb_2S_3 crystals. For NH_4Cl solutions the
growth rate of the faces also increases with rising temperature
in accordance with the inequality $v_{300°C} < v_{400°C}$.

　　3. On the basis of certain possible forms of Sb_2S_3 transport
in NH_4Cl, NH_4Br and Na_2S solutions, we have proposed a growth
mechanism for Sb_2S_3 crystals in both acid and alkaline media.

Literature Cited

1.　　R. A. Fisher, Statistical Methods of Investigation [in Russian], Gosstatizdat (1958).
2.　　Int. Assoc. Vulcanology, Int. Symp. Vulcanology, Wellington, New Zealand
　　　(1965), p. 45.

Crystallization in Systems $PbO-TiO_2-KF-H_2O$ and $PbO-ZrO_2-KF-H_2O$

V. A. Kuznetsov

In a previous communication [1] we considered the crystallization of titanium, zirconium, and hafnium oxides under hydrothermal conditions. In the present case we shall consider the crystallization of more complicated compounds containing titanium and zirconium. The principal subjects for consideration will be the systems $PbO-TiO_2-KF-H_2O$ and $PbO-ZrO_2-KF-H_2O$, these being studied in order to determine the conditions governing the synthesis and recrystallization of lead metatitanate and metazirconate ($PbTiO_3$, $PbZrO_3$). The choice of KF solutions as media for the synthesis and recrystallization of $PbTiO_3$ and $PbZrO_3$ was based on the results of our previous communication [1], in which* we demonstrated the possibility of transporting titanium and zirconium by these solutions. However, we also studied a number of other solvents of the kind most frequently used in hydrothermal investigations for the case of the $PbO-TiO_2-X-H_2O$ system (X = solvent): alkalis, chlorides, sulfates, borates, and so on. We found that in the quaternary $PbO-TiO_2-X-H_2O$ system titanium behaved very much the same as in the TiO_2-X-H_2O ternary system (formation of alkali titanates in solutions of sodium and potassium carbonates

---

*We know of only one previous paper on the hydrothermal synthesis of lead metatitanate [2]; $PbTiO_3$ single crystals were obtained in pure water at 380-450°C and at a pressure of 300-500 atm. As regards $PbZrO_3$, it was mentioned earlier [2, 3] that at high temperatures and water-vapor pressures lead metazirconate decomposed into PbO and ZrO_2. According to F. Jona and D. Shirane [4], repeated attempts have been made at producing titanate crystals (chiefly $BaTiO_3$) under hydrothermal conditions, but the results have always been negative, and have not been published.

81

and hydroxides, low mobility in sulfate and borate solutions, etc.),
thus constituting the most inert component of the system. Only
in KF solutions was the recrystallization of $PbTiO_3$ achieved.
Analogous results were obtained in the $PbO-ZrO_2-X-H_2O$ system,
in which KF was also the most "suitable" solvent for the synthesis
and recrystallization of lead zirconate.*

In the present investigation, in addition to the synthesis and
recrystallization of lead metatitanate and metazirconate, we also
considered the conditions of formation of a number of accompanying
phases. However, the principal investigations were centered around
the parameters which enabled $PbTiO_3$ and $PbZrO_3$ to be obtained
in the form of single crystals. This is why we shall not be describing crystallization at temperatures below 450°C in concentrated
KF solutions, while a number of factors which are of importance
in obtaining accompanying phases will not be considered in any
detail. Most of our experiments were carried out at 450-700°C
under pressures of 800-3000 atm. The PbO/TiO_2 (ZrO_2) weight
ratio in the original mixture was varied from 1:1 to 5:1, the
concentration of the solution being 10 wt.% in most of the experiments. According to our observations, increasing the KF concentration to 30% produced no marked changes in phase formation in the systems studied. The autoclaves were provided with
silver linings in order to prevent the reduction of lead oxide to
lead. Experience showed that copper, titanium, and iron linings
were unsuitable for working with lead oxide.

Phase Formation in the
$PbO-TiO_2-KF-H_2O$ System

Under the conditions of our experiments the principal factor
affecting phase formation in the $PbO-TiO_2-KF-H_2O$ system is
the oxide ratio PbO/TiO_2 in the original charge. The temperature
and pressure have no very great influence on the range of
stability of the phases which we shall be discussing, simply increasing the velocity of their formation from the oxides.

For a PbO/TiO_2 ratio in the original charge equal to 1:1-2:1,
dark green crystals are formed as long prisms (Fig. 1). According

---

*Solutions of NH_4F, although providing a greater titanium and zirconium mobility
than KF, were not employed for synthesizing lead titanates and zirconates, since they
promoted the intensive reduction of lead to the metallic state from PbO.

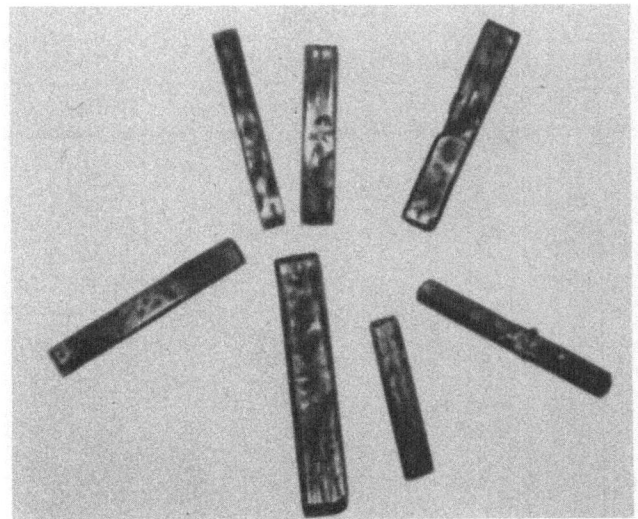

Fig. 1. Crystals of phase A (× 10).

to chemical analysis,* this phase (which we shall subsequently describe as phase A) contains PbO 47.71 and TiO$_2$ 50.70%, i.e., it constitutes the compound PbTi$_3$O$_7$. According to spectral analysis, fluorine is present (∼0.05%) in the crystals of phase A, together with traces of potassium and various impurities occurring in the solution. An x-ray diffraction recording of phase A is presented in numerical form in Table 1.

Crystals of phase A usually have a long prismatic habit and are formed as individual prisms 1-2 mm long. The crystals are transparent but usually deeply colored in green tones. Phase A is formed at all the temperatures and pressures studied, and its appearance is simply determined by the ratio of the oxides in the original charge. In the presence of a temperature gradient in the autoclave the crystals of phase A pass readily to the upper, "colder" zone of the reactor.

For a PbO/TiO$_2$ ratio (in the charge) of 1 : 1-3.5 : 1, crystals of phase B are formed. The maximum number of crystals of this

---

*The chemical analyses were carried out in the IGEM Laboratory by V. V. Danilova, G. A. Aralova, and Yu. S. Nesterova; the spectral analyses were carried out in the same laboratory by A. S. Dudykina.

TABLE 1

Phase A		Phase B		Phase C		Phase D	
d/n	I	d/n	I	d/n	I	d/n	I
3.52	3	3.96	1	3.95	1	4.03	1
3.23	10	3.78	4	2.97	5	3.70	1
2.975	10	3.21	6	2.85	6	3.37	5
2.92	3	2.98	3	2.76	10	3.11	2
2.876	1.5	2.85	3	2.33	1	2.63	3
2.63	3	2.76	1	2.22	1	2.49	10
2.46	1	2.58	10	2.02	2	2.21	1
2.337	2	2.31	2	1.95	2	2.19	2
2.23	1	2.13	2	1.84	3	1.80	1
2.158	9	2.10	4	1.78	3	1.55	2.5
1.90	3	2.085	3	1.74	1		
1.84	2	1.95	2	1.68	2		
1.739	3	1.904	3	1.56	2		
1.70	3	1.61	2	1.42	3		
1.62	7	1.592	3				
1.546	2	1.536	4				
1.481	1						
1.450	1						
1.400	1						
1.298	2						
1.234	3						

phase is obtained for an oxide ratio of $\sim 2:1$. The x-ray recording of the B-phase crystals is indicated in Table 1.

In external form the crystals of phase B are similar to those of phase A (Fig. 2), being dark green, flat needles up to 1-2 mm long, frequently forming fan-shaped aggregates. Usually phase-A crystals are formed together with phase B. This makes it difficult to separate the phases for chemical analysis, so that at the present time our knowledge of the composition of phase B is none too accurate: According to chemical analysis it contains TiO_2 72.2-72.82%, PbO 14%, and K_2O 11.2-12%. On this basis we may assign phase B the composition $K_2O \cdot 0.5PbO \cdot 7TiO_2$ (the main impurities being Na_2O up to 0.5 and F up to 0.3%).

Phase B is characterized by the greatest rate of formation of all compounds in the $PbO-TiO_2-KF-H_2O$ system which we synthesized. Thus, in experiments lasting 1-1.5 days with a charge

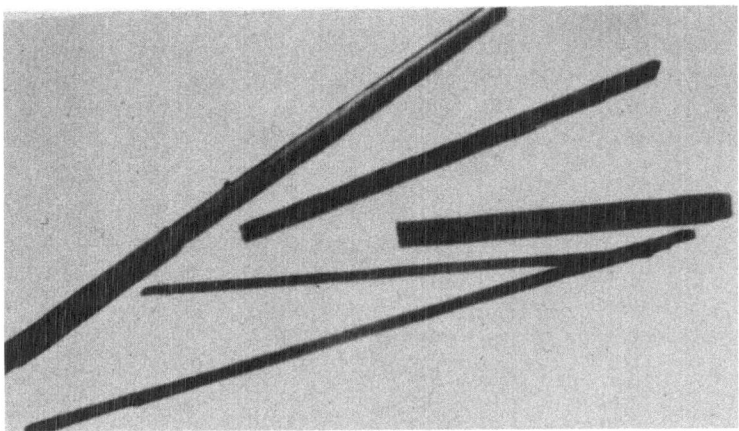

Fig. 2. Crystals of phase B (×20).

ratio PbO/TiO$_2$ of 1:1-4:1, crystals of phase B were always in
the majority of those formed. Under these conditions few crystals
of other phases were, as a rule, synthesized. On increasing the
period of the experiment, crystals of phase A and lead titanate
appeared in greater numbers; depending on the ratio of the components in the charge, these existed either in pure form or in association with one another in experiments of over 6-7 days duration. The metastability of phase B under the conditions of our
experiments is also indicated by the results obtained during its
recrystallization, when phase B partly "decomposed" into PbTi$_3$O$_7$
and PbTiO$_3$.

For a charge ratio of PbO/TiO$_2$ = 2.5:1-3.5:1, crystals of
lead metatitanate PbTiO$_3$ are synthesized in the form of cubes
(Fig. 3). The x-ray diffraction pattern of the resultant lead
metatitanate coincides with the standard x-ray diffraction pattern
of PbTiO$_3$ with a tetragonal cell.

For a high PbO/TiO$_2$ ratio in the charge (3.5:1 and over),
yellow lead oxide coexists with PbTiO$_3$, forming very fine "films."*

---

*According to [5], for a high content of PbO (over 83 mol.%) no independent com-
pounds are formed in the PbO–TiO$_2$ system, and there is only a solid solution of
PbTiO$_3$ in tetragonal lead oxide. The stability of the latter at room temperature
increases in the presence of titanium impurity.

Fig. 3. Lead metatitanate (× 5).

At the same time a number of phases, apparently of variable composition, are formed and intensively recrystallized. We shall consider two of these which are encountered most frequently.

Phase C is composed of long, white, transparent crystals (Fig. 4). According to chemical analysis, phase C contains PbO

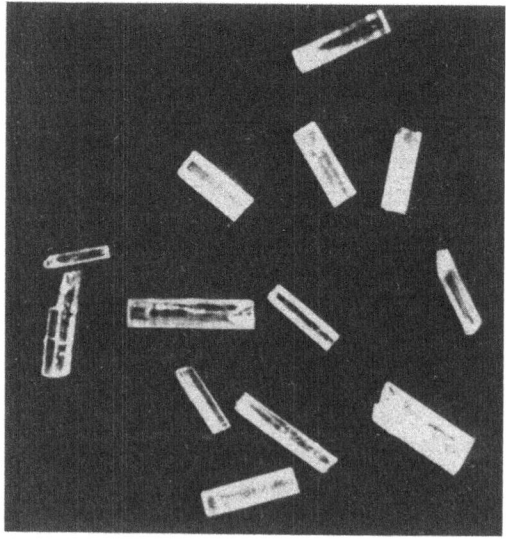

Fig. 4. Crystals of phase C (× 5).

91.4-93.55 and TiO$_2$ 0.5-3.68%, with between 0.29 and 0.49% of fluorine. Crystals of phase C evidently contain a variable amount of water, not taken into account in our analyses. An x-ray diffraction analysis of phase C is indicated in Table 1.

Under conditions analogous to those corresponding to the formation of the phase C, crystals of phase D may undergo synthesis and recrystallization in the form of white transparent hexagonal plates up to 2 mm in size (Fig. 5). According to chemical analysis, the D crystals have PbO content of 88%, with TiO$_2$ 0.56 and F 0.3%. The x-ray diffraction analysis of phase D is given in Table 1.

Phase Formation in the PbO–ZrO$_2$–KF–H$_2$O System

We studied this system under the same conditions as in the previous case; a smaller number of phases, however, was formed. Thus we found no compounds analogous to phases A and B in the system containing titanium oxide, only observing the zirconate PbZrO$_3$ and phase C. The PbZrO$_3$ is easily synthesized for a PbO/ZrO$_2$ ratio (in the charge) of $\sim$2:1, taking the form of fine cubic crystals. In Table 2 we compare the x-ray diffraction recording of our own synthesized lead zirconate with that of the standard compound. The x-ray diffraction pattern of the lead zirconate obtained corresponds to PbZrO$_3$ with a tetragonal cell.

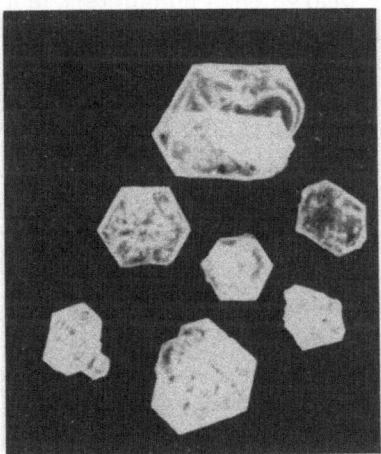

Fig. 5. Crystals of phase D ($\times$ 10).

TABLE 2

$PbZrO_3$ synthesized		$PbZrO_3$ standard		$PbZrO_3$ synthesized		$PbZrO_3$ standard	
d/n	I	d/n	I	d/n	I	d/n	I
4.15	4	4.16	50	1.85	2	1.86	30
2.92	10	2.94	100	1.68	5	1.70	100
2.38	3	2.40	60	1.45	2	1.46	85
2.06	3	2.08	70	1.37	1	1.39	50
		2.05	60	1.31	1	1.32	85

The crystals of phase C are formed for high PbO/ZrO_2 ratios (3 : 1 and over) and in external appearance (habit) are no different from the crystals synthesized in the $PbO-TiO_2-KF-H_2O$ system. The ZrO_2 content is 0.49-2.5%.

Synthesis and Recrystallization of $PbTiO_3$ and $PbZrO_3$

The synthesis of lead metatitanate from a mixture of the oxides in a 10% KF solution takes place at 450°C and ~1000 atm pressure. However, the velocity of the reaction under these conditions is quite low, and fine crystals of $PbTiO_3$ up to 0.3 mm in size are the result. On raising the temperature of the experiment the rate of synthesis increases, and at 700°C the reaction between the components of the charge ends completely in a few hours. The size of the crystals increases at the same time, in some cases to 1 mm. Larger crystals were obtained by the "divided charge" method, in which the original titanium oxide was placed in the lower zone of the reactor and the lead oxide in the upper. A temperature drop of 5-10°C was created between the zones, the rate of synthesis being controlled by regulating this. In this way $PbTiO_3$ crystals up to 2 mm in size were obtained in 20-h experiments at 700°C.

At temperatures above 500-550°C and in the presence of a temperature gradient between the upper and lower zones of the autoclave, as the synthesis of the $PbTiO_3$ from the oxides proceeded, recrystallization took place, and the compound passed into the upper zone of the reactor. Intensive recrystallization occurs when lead titanate previously synthesized either by the hydrothermal method or by sintering the oxides at 1000-1100°C for 1.5 h is used in the original charge. As a result of recrystallization, druses of PbTiO$_3$ crystals were formed in the upper zone of the autoclave (on the walls), in which the dimensions of individual crystals reached as much as 1 mm (for $\Delta T = 35°$). On reducing the temperature drop, the intensity of the mass transfer also diminishes, but the size of the individual crystals increases to 3 mm. The rate of growth of the (100) faces meanwhile increases to 0.2 mm/day.

If the upper zone of the autoclave contained seeds (PbTiO$_3$ crystals obtained by spontaneous crystallization), then these seeds started growing although the average growth rate of the (100) face at 600°C and $\Delta T \approx 35°$ was no greater than 0.05-0.07 mm/day in experiments lasting 2-5 days.

Increasing the concentration of the KF solution to 30% slightly increases the intensity of transport of the lead metatitanate, although the resultant crystals are more defective, being opaque, with a large number of inclusions. It is an interesting fact that in the presence of PbO no titanium fluorides are formed in very concentrated KF solutions, such as occurred in the TiO$_2$−KF−H$_2$O system.

The synthesis of lead metatitanate took place under the same conditions as that of PbTiO$_3$, although the velocity of the reaction was slightly lower and the resultant PbZrO$_3$ crystals were no larger than 0.5-1.0 mm. The recrystallization of PbZrO$_3$ also took place less intensively than that of PbTiO$_3$ under the same conditions. It was a typical occurrence that hardly any transport of ZrO$_2$ occurred in the KF solutions [1], the recrystallization of the PbZrO$_3$ only being explicable in terms of the action of the "extra" component, PbO.

The lead metatitanate and metazirconate crystals obtained in the experiments just described take the form of cubes. The PbTiO$_3$ crystals are transparent but yellowish. The crystals usually have a complex, many-domained structure; they contain inclusions of the mother solution in the center and are often cracked. Fine samples (up to 1 mm) are free from these defects, as well as crystals grown very slowly (the growth rate of the (100) face no more than 0.1 mm/day).

The $PbZrO_3$ crystals are usually cloudier, except for the very fine samples (~ 0.5 mm), and have a stronger yellow color.

Isomorphism of Titanium and Zirconium

It is well known that under natural conditions the isomorphism of titanium and zirconium only appears in exceptional cases. When studying the TiO_2-ZrO_2 system in [6] it was also established that only limited solid solutions were formed between the components. In the $PbTiO_3-PbZrO_3$ system [7] limited isomorphism between titanium and zirconium is found at 800-1200°C.

In order to estimate the isomorphism of Ti and Zr under the conditions of our experiments, we recrystallized material of composition $Pb(Ti_{0.5}Zr_{0.5})O_3$ in 10% KF solutions at 580-600°C. In the upper zone of the autoclave crystals of cubic form up to 1 mm in size were formed; according to x-ray data these were lead metazirconate. According to chemical analysis, the amount of TiO_2 present was about 25%.

Analysis of the charge after the recrystallization of $Pb \cdot (Ti_{0.5}Zr_{0.5})O_3$ showed that it contained crystals of phase C, zirconium oxide, and lead metatitanate.

Conclusion

In the hydrothermal growth of lead titanates and zirconates the titanium oxide and zirconium oxide constitute inert, low-mobility components in the majority of solvents, and the range of recrystallization of the $PbTiO_3$ and $PbZrO_3$ is limited by the conditions of high mobility of TiO_2 and ZrO_2. The most favorable media for the end in view are potassium fluoride solutions operating at high temperatures (over 550°C). The presence of an "additional" component, lead oxide, may exert a considerable mineralizing influence on the transport of TiO_2 and ZrO_2, this appearing particularly in the case of lead metazirconate.

Large $PbTiO_3$ and $PbZrO_3$ crystals may be obtained under hydrothermal conditions, either by controlled synthesis from the oxides (the "divided-charge" method being most promising) or by recrystallization in the presence of a temperature drop. The latter method may also be used for producing crystals of $Pb(Ti,$

Zr)O$_3$ solid solutions involving the mutual substitution of the titanium and zirconium over a fairly wide range.

Literature Cited

1. V. A. Kuznetsov, this volume, p. 43.
2. A. Christensen and S. Rasmussen, Acta Chem. Scand., Vol. 17, No. 3 (1963).
3. L. Reed and G. Katz, J. Amer. Ceram. Soc., 39:260 (1956).
4. F. Jona and D. Shirane, Ferroelectric Crystals [Russian translation], Izd. Mir (1965), p. 423.
5. Y. Matsuo and H. Sasaki, J. Amer. Ceram. Soc., 46:409 (1963).
6. N. A. Toropov, V. P. Barzakovskii, V. V. Lapin, and N. N. Kurtseva, Phase Diagrams of the Silicate Systems, No. 1 [in Russian], Izd. Nauka (1965), p. 280.
7. S. Fushimi and T. Ikeda, Japan, J. Appl. Phys., 3:171 (1964).

Growth Characteristics of Calcite Crystals in Aqueous Solutions of Carbonic Acid

N. Yu. Ikornikova

Natural crystals of Iceland spar are formed in both chloride and carbonate hydrothermal solutions. In this communication we shall present the results of some experiments on the growth of calcite crystals in aqueous solutions of carbonic acid. In relation to the composition of the virtual components* these solutions are analogous to natural carbonate thermal springs.

1. Structure and Properties of H_2O-CO_2 Solutions

The H_2O-CO_2 system has by now been studied in fair detail over a wide range of temperatures, pressures, and CO_2 concentrations. According to [1, 2], at pressures up to 60 bar the isotherms (from 10 to 50°C) represent $(H_2O)_l-(CO_2)_g$ equilibrium curves. These conditions are characterized by a rapid increase in the solubility of CO_2 with increasing pressure. In the range 100-500 bar the same isotherms correspond to the $(H_2O)_l-(CO_2)_g$ equilibrium, i.e., to conditions under which the solubility of CO_2 changes very little with pressure. In the temperature range 50-100°C and at pressures of up to 700 bar the system corresponds to the $(H_2O)_l-(CO_2)_g$ equilibrium, but in this case there is a gradual rise in the solubility of the CO_2 in water as pressure increases. In the temperature range 10-50° and the pressure range 1-700 bar, the solubility of CO_2 in water always rises with increasing pressure (T = const) and falls with increasing temperature (P = const).

---

*The components which enter into the formula of the phase rule.

In the same pressure range but at higher temperatures, the isotherms intersect, and the sign of the temperature coefficient of the
solubility of CO_2 reverses (in the range 50-100° and at P = 500
kg/cm^2). For temperatures of over 100° the isothermal curves
intersect in the pressure range 125-200 kg/cm^2 [3].

The complete diagram of the isotherms of the H_2O-CO_2
system is presented in Fig. 1 for the temperature range 100-
350°C, the pressure range 10-1600 bar, and the concentration
range 0-100 mol.% CO_2 [4-6]. The isotherms have the form of
loops, this being a characteristic of liquid—gas equilibrium con-

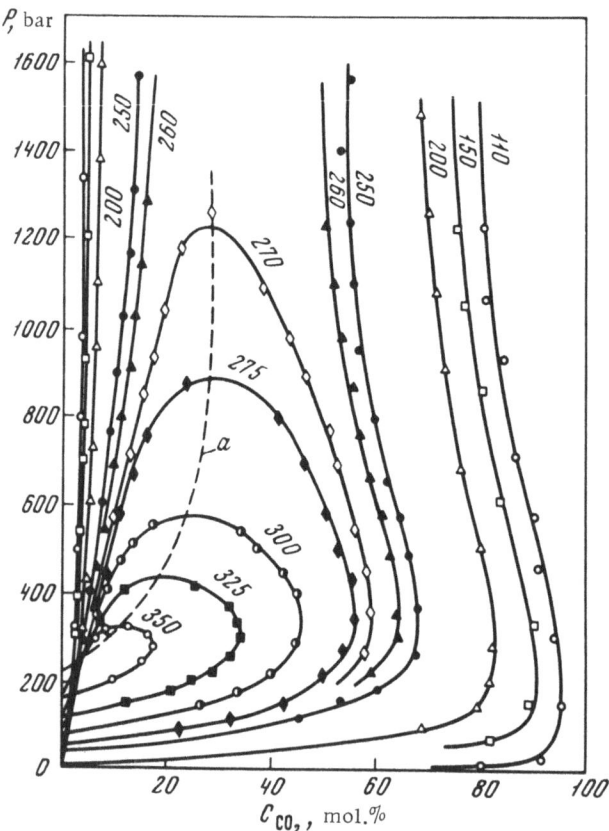

Fig. 1. Diagram of state (phase diagram) of the H_2O-CO_2 system, according to [4-6].

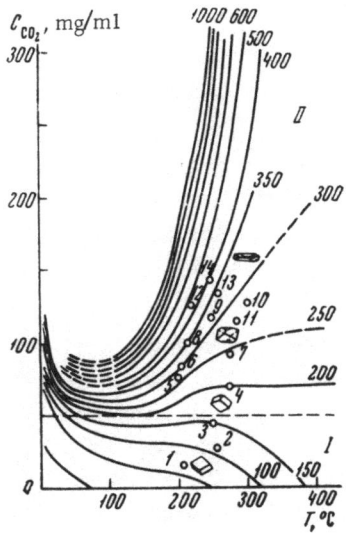

Fig. 2. Isobars of the H_2O-CO_2 system — our own construction based on an interpolation of data taken from [4-6]. The changes taking place in the habit of the calcite crystals with changing parameters are illustrated schematically as a background to the diagram; the numbers on the points represent the numbers of the experiments indicated in Table 3; the numbers on the curves give the pressure in bars; C_{CO_2} is the original concentration.

ditions. The axial line of the loops (a) is the critical curve. To the left of this is the region of curves characterizing aqueous solutions of CO_2, the H_2O component being predominant. These solutions are usually employed in hydrothermal experiments and correspond to the composition of natural carbonate thermal springs. The right-hand side of the diagram is the region of the gas phase. As already noted, the signs of the temperature coefficient of the solubility of CO_2 in water are opposite in the low- and high-pressure regions. Generally speaking, the H_2O-CO_2 system has been studied up to 3500 bar [7], but we shall not consider this part of the diagram here.

The change in the sign of the temperature coefficient of CO_2 solubility in water is demonstrated in the isobaric diagram which we have plotted on the basis of the data presented in [4-6] (Fig. 2). In this diagram we find two regions, I and II, lying on the two sides of the 200 bar isobar, which is close to the isobar of the critical pressure of water. In the temperature range 10-400°C the solubility of CO_2 in water falls with temperature in region I. In the pressure range 150-200 bar the CO_2 concentration varies little with temperature if the pressure is constant. In region II the isobars bend and approach the vertical asymptote more and more as the pressure increases:

$$\left| \frac{\Delta P_{CO_2}}{\Delta t} \right|_P \to \infty .$$

The diagram under consideration has yet another characteristic:
All the isobars above 150 bar exhibit a minimum between 100
and 150°. For isobars 500-1000 this minimum is shown by a
broken line and its position can only be presumed, as no experimental measurements have been made in this region.

Comparing the diagram under consideration with the data of
Ellis [8] we see clearly that the maximum of the Henry constants
also corresponds to temperatures of ~150°C. At the same temperature there is a change in the sign of the enthalpy ΔH_i. It
follows that in the temperature range 10-150° and for pressures
of up to 200 bar solutions of CO_2 in water obey the Henry law entirely, and constitute a typical gas–liquid system. At pressures
of over 200 bar and temperatures of over 150° there is a region of
homogeneous liquid phase, the properties and constitution of which
differ from those of aqueous solutions of carbon dioxide at pressures below 200 bar.

Of great importance in connection with the progress of the
processes taking place in the $H_2O–CO_2$ system is the pH/pressure
relationship. According to Moore [9] there is a tendency for pH
to fall with increasing P_{CO_2} under standard conditions. This may
explained by the fact that, with increasing CO_2 solubility, the concentration of both the hydrocarbonate and the hydrogen ions increases. At the same time the results of our own measurements*
of residual pH values at high temperatures and pressures (200-
600 bar and 230 and 300°C) showed that both CO_2 concentration
and pH value increased in solutions held for 24 h. Our method was
as follows. At the end of each experiment the heated autoclave
was cooled by quenching, after which the solution was sampled
through a high-pressure valve. In the extracted solutions we
measured the residual values of the pH factor and CO_2 concentration (titration with alkali) in parallel samples — one of these immediately after extraction (within 5-7 min) and the other after
holding in a hermetically closed vessel for 2 h. The results of

---

*The measurements were made in the apparatus illustrated in Fig. 4.

the measurements are given in Table 1. If the solutions are titrated immediately after extraction, then the concentration of the free carbonic acid is insignificant; however, it increases sharply if the solutions are titrated after a considerable period. Evidently the increase in the CO_2 concentration in solution takes place as the compounds formed at high pressures (which are unstable under standard conditions) decompose. We may suppose that in these compounds the CO_2 enters into the composition of complex molecules.

Thus regions I and II in Fig. 2 differ not only in the signs of the temperature coefficient of CO_2 solubility but also in respect of the relationship between pH and P_{CO_2}. The different properties thus appearing in the same system may clearly be explained by differences in the constitution of the liquid.

TABLE 1. Change in the pH Value (Residual) with Changing Pressure

According to [8]				Author's data							
0°		25°		230°				300°			
P_{CO_2}, atm	pH	P_{CO_2}, atm	pH	P_{tot}, kg/cm²	pH	C_{CO_2},mg/ml after 5-7 min	after 2 h	P_{tot}, kg/cm²	pH	C_{CO_2},mg/ml after 5-7 min	after 2 h
1.0	3.5	1.0	3.7	200	3.5—3.8	33	47	250	4.0	38	75
1.4	3.4	1.7	3.5	205	3.7	32	56	250	3.9	41	69
2.6	3.3	2.5	3.4	300	6.2	80	80	350	7.3	137	153
3.6	3.3	2.9	3.4	400	9.0	80	105	—	—	—	—
8.3	3.3	3.7	3.4	600	10.7	97	120	500	8.1	137	220
15.3	3.2	3.8	3.4								
23.4	3.2	5.4	3.3								
		5.8	3.3								
		7.2	3.3								
		7.8	3.3								
		9.5	3.3								
		10.5	3.3								
		12.7	3.3								
		18.7	3.3								
		38.3	3.3								

For fairly low pressures (up to 200 kg/cm²) aqueous solutions of CO_2 comprise a mixture of tetrahedral groups of water
molecules and O = C = O "dumbbells." It is hard to say how the
H_2O and CO_2 molecules are ordered with respect to one another.
In any case they cannot form strong bonds, since the CO_2 is quite
mobile, as indicated by the reduction in its concentration which
takes place with rising temperature.

At high pressure the CO_2 concentration reaches 4–5 wt.%
(40–60 mg/ml) and the short-range order in the structure of the
water is disrupted; the water molecules and dumbbells are re-
oriented and form compounds with more or less strong bonds.
In this region, as the pressure rises the CO_2 and HCO_3^- concentrations fall relative to the original and that of the hydroxyl ions
increases. Thus the upper part of the diagram relates to solutions
with evidently consist chiefly of molecules formed by the synthesis of water and free carbonic acid. The synthesis of the molecules
is accompanied by the evolution of hydroxyl ions and their ac-
cumulation in the system. Molecules of the formaldehyde type,
carboxyl acids, and intermediate products of these compounds are
probably formed from the water molecules and dumbbells.

We may reasonably assume the following reactions for the
synthesis of formic acid and formaldehyde molecules:

$$2H_2O + CO_2 + 2\bar{e} \rightleftharpoons HCOOH + 2OH^-,$$
$$3H_2O + CO_2 + 4\bar{e} \rightleftharpoons HCOH + 4OH^-.$$

Later it will be shown how the constitution of the solutions
is reflected in the habit of the calcite crystals.

2. Influence of P_{CO_2} and the Total
Pressure on the Habit of Calcite
Crystals

A number of investigations into water–carbon dioxide–carbonate systems over a wide range of P, T, and C have been published. Almost "dry" systems, containing no more than 20% of
H_2O, and essentially constituting solutions of water in a calcium
carbonate melt, were studied in [10] in connection with the problem
of the origin of carbonatites in the temperature range 500–1300°C
and the pressure range 27–300 atm. In aqueous solutions of carbonic acid in which the $H_2O + CO_2$ constitutes at least 80–90% only

two solid phases may be formed – calcite $CaCO_3$ and portlandite $Ca(OH)_2$ [11]. At room temperatures portlandite is found even for low values of P_{CO_2} of the order of minus 12 atm [12].

The solubility of calcite in aqueous solutions of carbon dioxide was studied between 10–300°C and 1–100 bar in [3, 13–16]. According to these data, for P = const the solubility falls exponentially with rising temperature. At the same time for T = const the solubility rises exponentially with increasing P_{CO_2}. Later experiments [17] at 200–600°C and 200–1400 bar in solutions of 10–100 mg/ml CO_2 showed that the absolute solubilities of calcite at high temperatures and pressures were almost an order of magnitude lower than those measured at pressures up to 100 bar. The laws governing the changes taking place in the solubility of calcite at high temperatures and pressures were the same as at low, i.e., for constant total pressure the solubility falls with rising temperature, while it increases with rising total pressure. Thus over a wide range of temperatures and pressures the temperature coefficient of the solubility of calcite has a negative sign.

Special attention should be drawn to the extremal values found on the calcite solubility isobars (Fig. 3). The jump on the isobars occurs at the same solvent concentrations (40–60 mg CO_2 per ml H_2O) as those at which the hypothetical boundary between regions I and II appears in Fig. 2; hence the isoconcentration line denoted by the broken horizontal line in Fig. 2 is in fact the

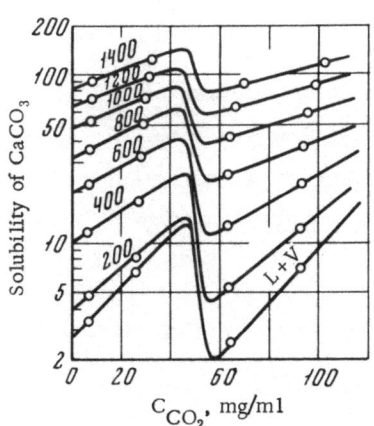

Fig. 3. Isobars of the solubility of calcite in aqueous solutions of carbonic acid [17] at T = 300°C. The figures on the curves give the pressure in bars.

boundary between regions with different structures (constitutions) of the liquid.

Thus analysis of H_2O-CO_2 and $H_2O-CO_2-CaCO_3$ systems indicates the existence of aqueous solutions of carbon dioxide with different structures.

So far as we know, none of the publications relating to the solubility of calcite in aqueous solutions of carbonic acid have mentioned the formation of portlandite. In our own experiments, not only was this compound observed, but its role in the morphological changes of the calcium carbonate crystal was also indicated.

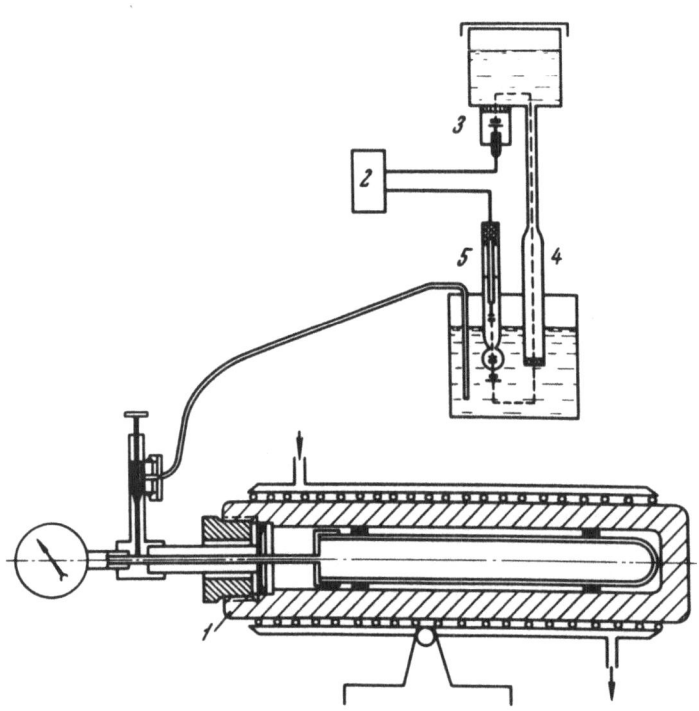

Fig. 4. Arrangement of the apparatus for synthesizing calcite crystals in aqueous solutions of carbonic acid under isothermal conditions. 1) Autoclave with glass or Teflon linings; 2) LIM-60M pH meter; 3) portable auxiliary silver chloride electrode; 4) electrical contact (key, switch); 5) electrode made of lithium glass (indicator); this apparatus was used to measure the residual pH values in H_2O-CO_2 solutions.

TABLE 2. Results of the Chemical Analysis of the Solid
Phase Formed in the $H_2O-CO_2-CaCO_3$ System

Solid phase	Comp., %	P, kg/cm²	T, °C	Solid phase	Comp., %	P, kg/cm²	T, °C
$CaCO_3$	99.98	70	200	$CaCO_3$	97.71	275	300
$Ca(OH)_2$	—			$Ca(OH)_2$	0.79		
Insoluble residue (i.r.)	0.02			i.r.	1.50		
				$CaCO_3$	97.28	270	220
$CaCO_3$	98.30	200	275	$Ca(OH)_2$	1.30		
$Ca(OH)_2$	—			i.r.	1.42		
i.r.	1.70			$CaCO_3$	97.15	330	250
				$Ca(OH)_2$	1.65		
				i.r.	1.20		

In order to grow calcite crystals (at the temperature and
pressure denoted by the points in Fig. 2) we used an isothermal
method in which the crystal grew with changing CO_2 concentration as
the pressure was reduced at T = const. The charge was dissolved
in a horizontal, rocking autoclave until equilibrium was achieved
(Fig. 4), after which the pressure was slowly released through
a valve.

The solid phases obtained in the $CaCO_3-CO_2-H_2O$ system
(containing 0.5 N NaCl) at 200-300°C and 70-500 kg/cm² in the
CO_2 concentration range 12-150 mg/ml constituted well-formed
crystals of calcite and portlandite scales (flakes) (Table 2). The
portlandite was recognized in several ways, depending on the
amount present: by chemical analysis, by x-ray spectroscopy, and
(in cases in which it occurred as individual flakes) by crystal-
optical methods (Table 3).

At pressures of under 150 kg/cm² rhombohedral calcite
crystals without any pinacoid are formed. The pinacoid appears
on crystals formed at pressures* of about 200 kg/cm², and although
portlandite is not found in this part of the system the appearance
of the pinacoid indicates initial changes in the concentrations of
the components of the solution. Starting from 200 kg/cm², as the

---

*That is, under the conditions in which the boundary between regions with different
signs of the temperature coefficient of solubility of carbon dioxide appears on the H_2O-
CO_2 equilibrium diagram.

TABLE 3. Results of Experiments Aimed at Synthesizing Calcite Crystals in $H_2O-CO_2-0.5$ N NaCl Solutions

Experiments (see Fig. 2)	T, °C	P, kg/cm²	$c_{Ca(OH)_2}$ in solid phase	Method of recognition	Characterisitics of CaCO₃ crystals	
					shape	dimensional ratio along L_3^6 and L_2
1	202	70	—	By none of the methods was Ca(OH)₂ detected	Rhombohedron (10Ī1)	1:1
2	260	110				
3	250	150			Rhombohedron (10Ī1) with barely perceptible (000Ī) truncation	0.9:1
4	275	200	Individual plates of portlandite	Crystal-optical, Debye photographs		
5	200	270			(10Ī1), habit (0001), subordinate	0.8:1
6	215	300	0.50	Crystal-optical	(10Ī1) and (0001) similar in dimensions	4:5
7	275	240				
8	225	350	1.30		(10Ī1) and (0001) uniformly developed	3:5
9	250	320	1.11			
10	290	277	1.32	Quantitative analysis of the whole solid phase and x-ray monitoring of the phases	(0001) greatly predominates	1:3
11	300	275	1.25			
12	220	470	1.70		Crystals of plate-like habit, (0001) predominates, (10Ī1) subordinate	1:6
13	265	340	2.25			
14	250	390	2.29			1:8

total pressure and CO₂ concentration increase and as OH⁻ ions accumulate in the system, the pinacoid develops more and more on the calcite crystals and the amount of Ca(OH)₂ in the solid phase becomes greater (Fig. 2). The portlandite* is deposited in the

---

*Portlandite is a rare and little-known mineral; it is encountered in one contact-metasomatic site in Antrim (Ireland) in association with a hydrated compound of calcium carbonate and silicate at small depths. The mineral has been obtained artificially a number of times when studying calcium carbonate systems, and its structure and optical properties have been determined. Trigonal (rhombohedral) portlandite (D_{3d}^3) forms colorless plates along the (0001) with a characteristic mother-of-pearl luster. The crystals also show the faces of the prism and one of the rhombohedra with $\rho = 34°$.

form of a fine-sclaed powder covering the surface of the calcite crystals. The change in the habit of the calcite crystals associated with the formation of $Ca(OH)_2$ in the system may easily be explained if we compare the structures of the two phases (Fig. 5) [18, 19].

The structure of portlandite consists of hexagonal layers in which all the octahedra are occupied by cations, while in the structure of calcite $^1/_3$ of the octahedra are populated in each horizontal layer, and in each of three contiguous octahedral columns two empty and one populated octahedron alternate. In both structures calcium cations occur in an octahedral coordination with six oxygen atoms (calcite) or OH^- groups (portlandite) around them. In the calcite structure the carbon atoms are situated within a plane triangle which serves as a common face for each pair of empty octahedra. As the concentration of OH^- groups increases at the expense of a reduction in CO_2 and HCO_3^- content, the portlandite structure is built; on the other hand, as the CO_2 and HCO_3^- ion concentrations increase the calcite lattice is formed.

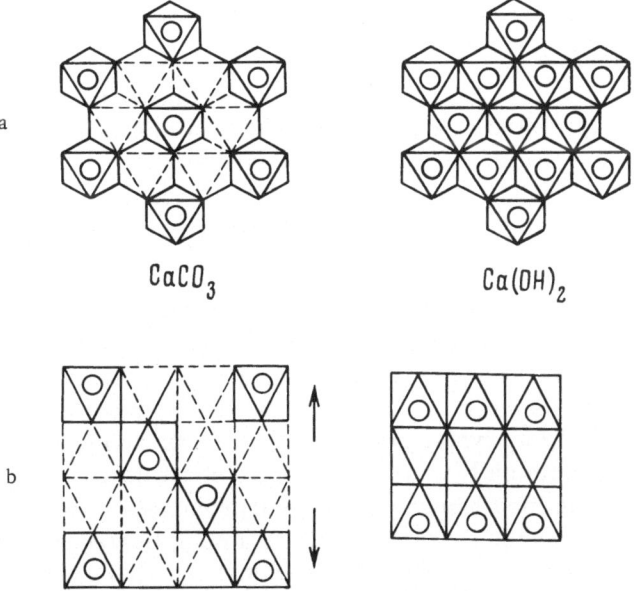

Fig. 5. Structure of calcite and portlandite in projections on the dense-packing plane (a) and on a plane parallel to the $(10\bar{1}0)$ (b).

As free hydroxyl ions accumulate in the solution, some of the calcium ions are joined to them, forming portlandite groups (associates) in the liquid phase. Since the two structures are mutually interchangeable, as the developing $Ca(OH)_2$ associated approach the surface of the calcite the CO_3 groups may be replaced by hydroxyls.

Orienting themselves along the calcite (0001) face, the associates form adsorbed layers and retard the diffusion of the solution to this plane. The retarded diffusion in the [0001] direction changes the relative growth rates of the faces, reducing the growth rate of the pinacoid. Earlier we showed how a change in the concentration ratio of the components of the system affected the evolution of the habit from rhombohedral to pinacoidal.

We shall now consider the role of impurities in the development of the rhombohedral form of the crystal.

3. Formation of the Rhombohedron

The development of the rhombohedral form of the crystals may be observed during crystallization, first in pure solutions, and then on introducing impurity (NaCl) with a gradually increasing concentration.

In this series of experiments the crystals were grown in the apparatus illustrated in Fig. 4 but in Teflon linings. We used a specially pure $CaCO_3$ reagent (for chromatography), twice-distilled water, and solid CO_2. Chloride impurities were completely absent from the charge and solvent, while the content of Group II elements amounted to thousandths of a percent.

In pure carbon dioxide solutions not containing any traces of chloride impurities, crystals exhibiting a poor ability to develop characteristic faces are formed. Crystal aggregates and druses form on the walls on the lining. Crystals grow in the druses in the form of single-crystal grains with complex surfaces. Faces are not noticeable on the surface of these crystals ($\times 500$). Certain crystals are flat and sometimes have approximate hexagonal features; their surface usually exhibits a complicated relief. One of these crystals is shown in Fig. 6. Cleavages have been formed in the crystal, and it is quite evident that this particular sample is a trigonal single crystal in which L_3^6 is perpendicular to the plane of the photograph.

Fig. 6. Calcite crystal formed in H_2O-CO_2 solutions
containing no chlorides — it is distinguished by the
absence of faces. The clear rhombohedral cleavages
indicate that the sample is a single crystal with a surface having a complex relief.

In the aggregates, crystallization is accompanied by the
enlargement of some of the crystals at the expense of others; this
is evidently due to the dependence of the solubility of a crystalline
individual on its diameter. According to the Ostwald—Freundlich
rule these quantities are inversely related [20]. The aggregates
easily break down into individual single-crystal grains up to 6-8
mm in size.

In subsequent experiments, the proportion of NaCl was
gradually increased (from 0.05 to 0.5 N). In the presence of an
insignificant amount of NaCl, (10$\bar{1}$1) planes appeared on the crystals. There was a clear relationship between the amount of
chloride in the H_2O-CO_2 solution and the development of rhombohedral faces on the crystal. If a series of experiments is carried
out at exactly the same temperature, and in each case the same
total pressure is created (constant pressure while equilibrium is
maintained), but the NaCl content is varied (increased gradually),
then areas of (10$\bar{1}$1) faces appear on the surface of the crystal

TABLE 4. Development of the (1011)
Planes on the Surface of a Crystal in
Relation to Impurity Content

Concentration (N)	Number of flat $(10\bar{1}1)$ regions on the surface of the grain	Dimensions of the flat $(10\bar{1}1)$ regions on the grain surface, mm
0.05	Flat parts of individual grains	0.05—0.09
0.1	3—5	0.5—1
0.2	2—3	1—2
0.3	5—7	1—2
0.4	3—6	3—4
0.5	Rhombohedron $(10\bar{1}1$	

grains and gradually increase in size. We made a statistical computation of these flat areas of the $(10\bar{1}1)$ type, their numbers and dimensions, in relation to the NaCl concentration (Table 4).

Only on a few single crystals grown in $H_2O—CO_2—0.5N$ NaCl solutions did we observe microscopic areas of one of the rhombohedral faces, separated by a complex spherical surface. In the same solutions (but containing 0.5 N NaCl) all the crystals are formed with $(10\bar{1}1)$ faces.

When sodium chloride is introduced into the system, the CO_2 concentration changes at the same time as the NaCl concentration. The solubility of CO_2 in water diminishes as the chloride passes into the solution. This law may be illustrated by the temperature curves of the Henry constants K = f(T) and the Ostwald distribution coefficients λ = f(T) in solutions of $H_2O—CO_2$ and $H_2O—CO_2—$NaCl [8]. Thus

$$K_{\text{Henry}} = \frac{P_\alpha \quad \text{(partial pressure)}}{C \quad \text{(concentration)}}$$

in CO_2 solutions containing NaCl is higher than in the same solutions without any NaCl. At the same time

$$\lambda_{\text{Ostwald}} = \frac{C_{CO_2} \quad \text{in the liquid phase}}{C_{CO_2} \quad \text{in the gas phase}}$$

is lower in solutions of CO_2 containing NaCl.

Despite the fact that in the H_2O-CO_2 solutions the CO_2 concentration falls on introducing sodium chloride, the solubility of calcite (for P = const) increases in the presence of NaCl in aqueous solutions of carbon dioxide. For pressure of no greater than 200 atm the equilibrium curves of carbonate-water systems containing carbonic acid with and without NaCl obey the same hyperbolic law. Independently of whether the aqueous solution of carbonic acid contains sodium chloride or not,* the shape of the curve is the same, and the sign of the temperature coefficient of solubility remains constant. On changing the NaCl concentration (0−1 *M*) only the position of the curve relative to the axis changes, since the sodium chloride slightly increases the absolute values of the solubility.

The role of the impurity component arises from the fact that, as the chloride is adsorbed by the rhombohedral planes of the calcite, the rhombohedron gradually develops.

4. Antiskeleton

Crystals obtained in the same medium (H_2O-CO_2-n NaCl) but under conditions involving a temperature drop in a vertical autoclave have a peculiar morphology.

Under such conditions the presence of carbonic acid complicates convection. We observed this effect directly in glass autoclaves designed for a pressure of 10 kg/cm^2. The transparent autoclaves were made from thin-walled, annealed glass tubes (wall thickness up to 15 mm), closed with rubber obturators at the two ends, which were filled with water or weakly-concentrated NaCl solutions. In one half of the tubes, branched aragonite aggregates were placed, and in the other plates of calcite. A glass tap was sited in one of the rubber obturators, and CO_2 was introduced through this from a gas cylinder. We were able to observe the formation of two separate liquid phases at room temperature. After mixing (by rocking), some of the carbonic acid dissolved in the water while some remained in the form of a liquid layer under the aqueous solution. The CO_2/H_2O ratio varied in different experiments from 1:4 to 1:1. The temperature range in these experiments was 20-80°C. On heating part of the glass autoclave (that in which the calcite plates were placed) the forma-

---

*Up to 1 mole, as indicated in [8].

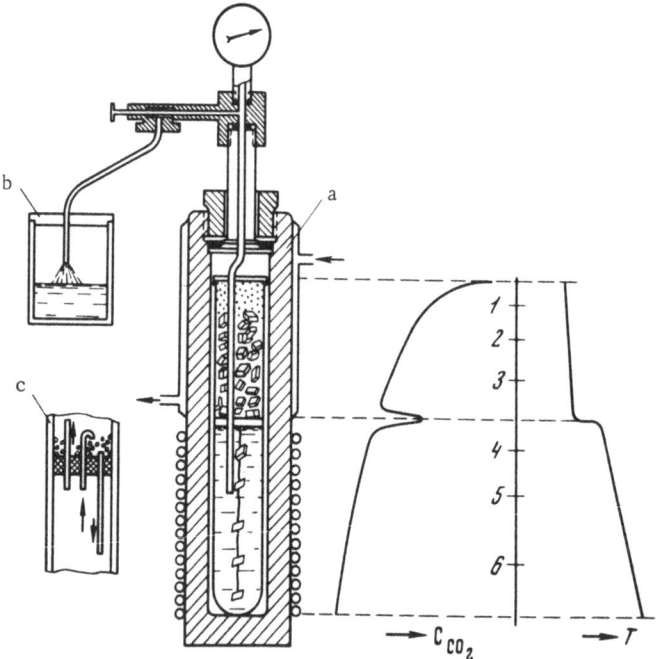

Fig. 7. Arrangement of the apparatus for growing crystals in aqueous solutions of CO_2 with a negative temperature coefficient of solubility, together with curves illustrating the height distribution of temperature and CO_2 concentration. a) Autoclave with glass linings; b) device for sampling; c) device for the convection of the solution saturated with CO_2 (barrier with tubes and a siphon); the numbers indicate the levels at which the solution samples were taken; analyses of the samples are given in Table 5.

tion of a gas phase was clearly observed. Thus under conditions involving a temperature drop convection takes place in the three-phase system.

This effect also occurs at higher temperatures and pressures when (under conditions involving a vertical temperature drop) the CO_2 concentration in the water changes sharply in zones of different temperature, and the rising convection currents predominate over the falling currents, thus creating conditions of considerable supersaturation in the lower zone of the autoclave. In the lower half of the autoclave shown in Fig. 7, moreover, a second liquid phase (liquid CO_2) is formed under the barrier, and this impedes convection.

In order to equalize the convection, we used a jet-directing device in the form of a reflux siphon (Fig. 7c). The experiments which were carried out in this autoclave were supplemented by chemical analysis of the solutions taken at high temperatures and pressures (Fig. 7b). The samples were extracted through a valve by means of glass tubes, which in each experiment were let down a prescribed distance into the inner vessel containing the solution. Each sample was taken twice at the same level — 5 h from the initiation of the process and 24 h after (Table 5). The curves drawn alongside the apparatus (Fig. 7) show that over the hot (crystallization) zone the CO_2 concentration gradually increases from the bottom toward the barrier, and close to the barrier itself a liquid layer of carbon dioxide is formed. At the same time, at each point of the solution the CO_2 concentration varies with time (i.e., it diminishes), and crystal growth takes place under conditions of supersaturation.

In the crystals formed in this system under conditions involving a temperature drop, with the autoclave in a vertical position, the rhombohedral faces are always covered with stepped pyramids; this occurs in parasitic crystals, in the overgrowth of rhombohedral seeds, and also in regenerated spheres. As the crystal grows, the flat area of the $(10\bar{1}1)$ face gradually diminishes, its tangential growth ceases and its growth along the normal remains intact. Thus the face ceases growing, but at the same time no other faces appear on the crystal such as occurs when the relative growth rates alter. In place of the face of the small rhombohedron $(\bar{1}101)$, a surface covered with bilateral accessories or "houses" formed by neighboring $(10\bar{1}1)$ faces (Fig. 8) develops. In place of the pinacoid,

TABLE 5. Analysis of an Aqueous Solution of Carbon Dioxide Under Conditions Involving a Temperature Drop

Sample (Fig. 7)	T, °C	Concentration of CO_2, %		Sample (Fig. 7)	T, °C	Concentration of CO_2, %	
		after 5 h	after 24 h			after 5 h	after 24 h
1	180	37	21	4	197	22	11
2	181	30	20	5	200	10	7
3	183	31	18	6	205	7.5	4

*Samples taken at different heights of the vessel containing the solution.

Fig. 8. Antiskeleton of the rhombohedron of calcite (a), and calcite rhombohedron
before its degeneration into an antiskeleton (b).

a surface covered with trilateral pyramids is formed. By virtue
of the development of the stepped structure on all six faces of the
rhombohedron, the crystals become lens-like in shape. When the
steps develop mainly on opposite parallel faces, the crystals are
reminiscent of a harmonic curve. When the tangential growth of
all the $(10\bar{1}1)$ faces stops uniformly, the crystal takes the shape
of a trefoil (clover). These shapes are called antiskeletons;
they are found in certain natural crystals, including natural calcite [20]. S. A. Stroitelev also observed these effects in artificial
crystals.

An antiskeleton constitutes a shape inverse to a skeleton;
its mode of genesis has been little studied, but it may be explained
by means of the inverse scheme of Humphrey-Owen [22], who indicated the disposition of the isoconcentration curves around the
surface of a crystal. According to this scheme, if the crystal
grows as a result of diffusion, the concentration of the solution
increases from the surface into the inside of the solution, and for
a large concentration drop a skeleton may be formed. The morphology of the antiskeleton is due to the inverse disposition of the
isoconcentration lines, in which the concentration reaches a
maximum at the surface of the crystal and falls on moving away
from the surface. In this way we may observe the degeneration

of the calcite rhombohedron into an antiskeletal shape under conditions of supersaturation.

* * *

In conclusion it should be noted that, in aqueous carbon dioxide solutions, the habit of calcite crystals changes on changing the concentration of the components, in the same way as we noted earlier for calcite crystals formed in aqueous chloride solutions [23].

The formation of a well-faced calcite crystal in the present system requires the presence of an impurity component. For the $(10\bar{1}1)$ shape to appear in the crystals, in particular, the solution must contain traces of a compound forming an associate with the NaCl type of structure. The change in the habit from rhombohedral to pinacoidal is associated with a change of free CO_2 concentration in the system.

Thus on the basis of the examples considered in the present communication we may readily appreciate the role of both the virtual and the impurity components in forming the crystal habit.

Literature Cited

1. R. Wiebe and J. Gaddy, J. Amer. Chem. Soc., 61:315 (1939).
2. R. Weibe and J. Gaddy, J. Amer. Chem. Soc., 63:475 (1941).
3. S. D. Malinin, in: Geochemical Investigations at High Temperatures and Pressures [in Russian], Izd. Nauka (1965), p. 48.
4. N. I. Khitarov and S. D. Malinin, Geokhimiya, No. 3, p. 18 (1956).
5. S. D. Malinin, Geokhimiya, No. 3, p. 292 (1959).
6. S. Takenouchi and G. C. Kennedy, Amer. J. Sci., 262:1055 (1964).
7. K. Todheide and E. U. Frank, Z. Phys. Chem., 37:387 (1963).
8. A. I. Ellis and R. Golding, Amer. J. Sci., 261:47 (1963).
9. M. Moore and J. Buchanan, J. Sci. Iowa State College, 4:431 (1930).
10. P. Wyllie and O. Tuttle, in: Questions of Theoretical and Experimental Topography [Russian translation], IL, (1963), p. 66.
11. P. J. Wyllie and E. F. Raynor, Amer. Mineral., 50:2077 (1965).
12. G. Grezes and M. Basset, Compt. Rend., 254:263 (1962); 260:869 (1965); 262:1217 (1966).
13. J. Miller, Amer. J. Sci., 250:161 (1952).
14. A. I. Ellis, Amer. J. Sci., 257:354 (1959).
15. A. I. Ellis, Amer. J. Sci., 261:259 (1963).
16. E. K. Seignit, H. D. Holland, and C. J. Biscardy, Geochim. Cosmochim. Acta, 26:1301 (1962).

17. W. E. Sharp and G. C. Kennedy, J. Geol., 73:40 (1965).
18. N. V. Belov, Structure of Ionic Crystals and Metallic Phases [in Russian], Izd. AN SSSR (1947).
19. H. Strunz, Mineralogical Tables [Russian translation], GNTI Lit. po Gornomu Delo, Moscow (1962), p. 43.
20. D. N. Grigor'ev, Ontogeny of Minerals [in Russian], Izd. L'vov. Univ. (1961), p. 203.
21. I. I. Shafranovskii, Crystals of Minerals, Curved-Faced Skeletal and Granular Shapes [in Russian], GNTI Lit. po Geol. i Okhrane Nedr., Moscow (1961), pp. 191, 193.
22. S. P. F. Humphrey-Owen, Proc. Roy. Soc., 197:218 (1949).
23. N. Yu. Ikornikova, in: Hydrothermal Synthesis of Crystals [in Russian], Izd. Nauka (1968), p. 114.

Dissociation Curves of Trigonal Carbonates

N. Yu. Ikornikova and V. M. Sheptunov

The limits to the existence of a carbonate crystal in a gas medium are defined by its dissociation and melting curves. In corrosive media (salt melts, aqueous salt solutions) these limits tend to occur at lower temperatures and pressures.

The diagram of equilibrium associations in the $MeO-CO_2$ system shown in Fig. 1 is typical of any trigonal carbonate. According to this diagram, the dissociation curve ends at the nonvariant point A, at which melting begins. From this point three curves of monovariant equilibrium diverge: AC, the curve corresponding to the decomposition of the liquid phase, AD, the incongruent melting curve of $MeCO_3$, and AE, the congruent melting curve of $MeCO_3$ (theoretically an isotherm). The diagram also shows the divariant fields, in which III and IV are the stability fields of $MeCO_3$.

According to Wyllie [1], an experimental diagram corresponding to this basic scheme can only be constructed for calcite. Analogous diagrams cannot be constructed for the other carbonates owing to the lack of experimental material; it was therefore decided to estimate the stability boundaries theoretically.

A theoretical dissociation curve (based on our own data) is presented in Fig. 2 together with the experimental melting and dissociation curves of calcite (2, 3).

The curve starts at the dissociation point for a pressure of 1 atm and ends at the point with parameters 1242°C and 44.11 atm, close to the experimental nonvariant point. A comparison

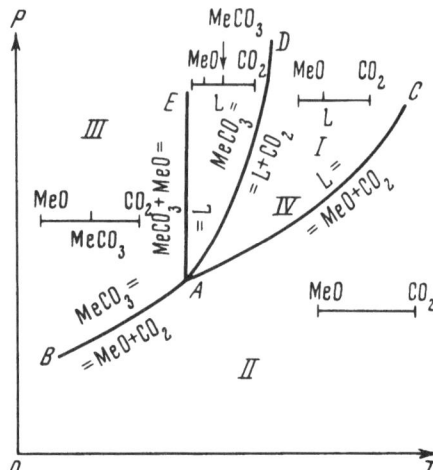

Fig. 1. Equilibrium associations in the
MeO–CO$_2$ system. I) - IV) Divariant
fields.

between the parameters of the experimental and calculated data
showed that, within the limits of experimental error, the calculated
and experimental dissociation curves of the solid material coin-
cided. This indicates that, over the range of parameters within
which dissociation occurs, ΔH changes very little (if at all) with
temperature and pressure. The section BD (Fig. 2) between the
experimental melting curves of calcite and the dissociation of the

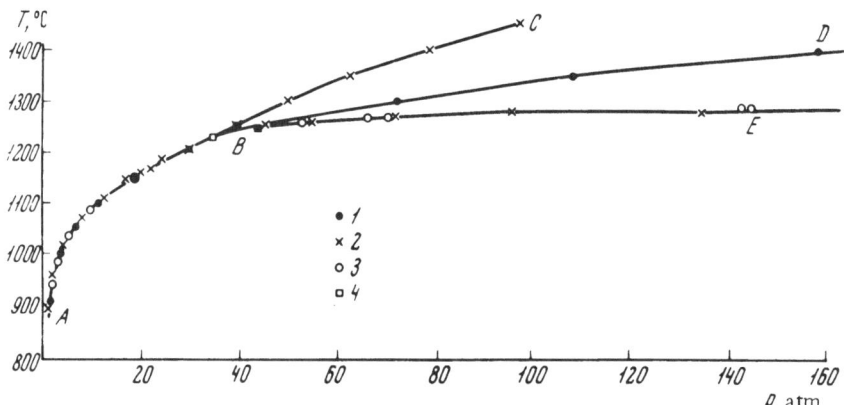

Fig. 2. Monovariant equilibrium curves of CaCO$_3$ = CaO + CO$_2$ (Tables 1 and 2). AB =
calculated dissociation curve of CaCO$_3$ coinciding with the experimental points: ● calculated (our own data); × experimental data [2]; ○ the same [3]; □, ■ nonvariant points;
BD = extrapolation of the calculated dissociation curve; BE = melting curve; BC = dissociation curve of the melt.

TABLE 1. Parameters of the Points on the Dissociation Curves of $CaCO_3$ in the Solid State According to Computed and Experimental Data (Fig. 2)

Our own calculated data		Experimental data of Baker [2]		Experimental data of Smith and Adams [3]	
T, °C	P, atm	T, °C	P, atm	T, °C	P, atm
900	1.00	900	1.00	906	1.0
960	2.02	959	2.30	942	2.0
1000	3.39	997	3.00	939	2.7
1050	6.23	1012	4.29	1039	4.9
1100	10.94	1071	7.94	1090	9.0
1150	18.45	1093	10.12	1150	18.7
1200	30.06	1107	11.95	1230	34.5
1242	44.11	1139	14.94	1233	40.2
		1147	17.00		
		1164	20.00		
		1163	21.80		
		1185	24.50		
		1212	30.00		
		1242	39.50		

TABLE 2. Parameters of the Points on the Melting Curve and the Dissociation Curve of the Melt According to Experimental Data (Fig. 2)

Experimental data of Baker				Experimental data of Smith and Adams (melting curve)	
melting curve		dissociation curve			
T, °C	P, atm	T, °C	P, atm	T, °C	P, atm
1250	45.5	1250	40.5	1269	53.00
1260	55.0	1300	50.0	1273	66.47
1270	72.0	1350	63.0	1275	71.00
1280	96.0	1400	79.0	1290	143.00
1290	135.0	1450	98.5	1289	144.00
1300	200.0				

melt corresponds to the equilibrium curve which would exist if the substance remained in the solid state for these parameters (Tables 1 and 2).

We also calculated the dissociation curves of other carbonates.

The method of computing the curves was based on the experimental determination of ΔH by a thermographic method. Derivatograms were plotted in a derivatograph of the L. and Y. Paulig type (ERDEY), using artificial single crystals. The samples were heated at 120°/min in platinum crucibles. The temperature was measured with a platinum–platinum-rhodium thermocouple. Heating leasted 100 min.

It is well known that the temperatures of the endothermic effects in different samples of the same material sometimes differ considerably if the samples have different particle sizes or contain different impurities.

We therefore studied the effect of grain size on the temperature of the endothermic effect. The discociation temperatures

TABLE 3. Dissociation Temperatures of Carbonate (in °C) Measured in Samples with Different Grain Sizes at P = 1 atm

Sample	T, °C		Tempera- ture dif- ference	$\Delta F°$
	sample size 0.5-1 mm	sample size 0.1-0.2 mm		
$CaCO_3$	910	9C0 900	10	−269.78
$MnCO_3$	543 545	527	16	−195.4
$CoCO_3$	485	452 453	33	155.57
$CdCO_3$	498 499	580	18	−160.2
$FeCO_3$	500	490	10	−161.06
$ZnCO_3$	460 460	—	—	−174.8
$MgCO_3$	550	⋯	⋯	−246
$NiCO_3$	500	—	—	−147.0

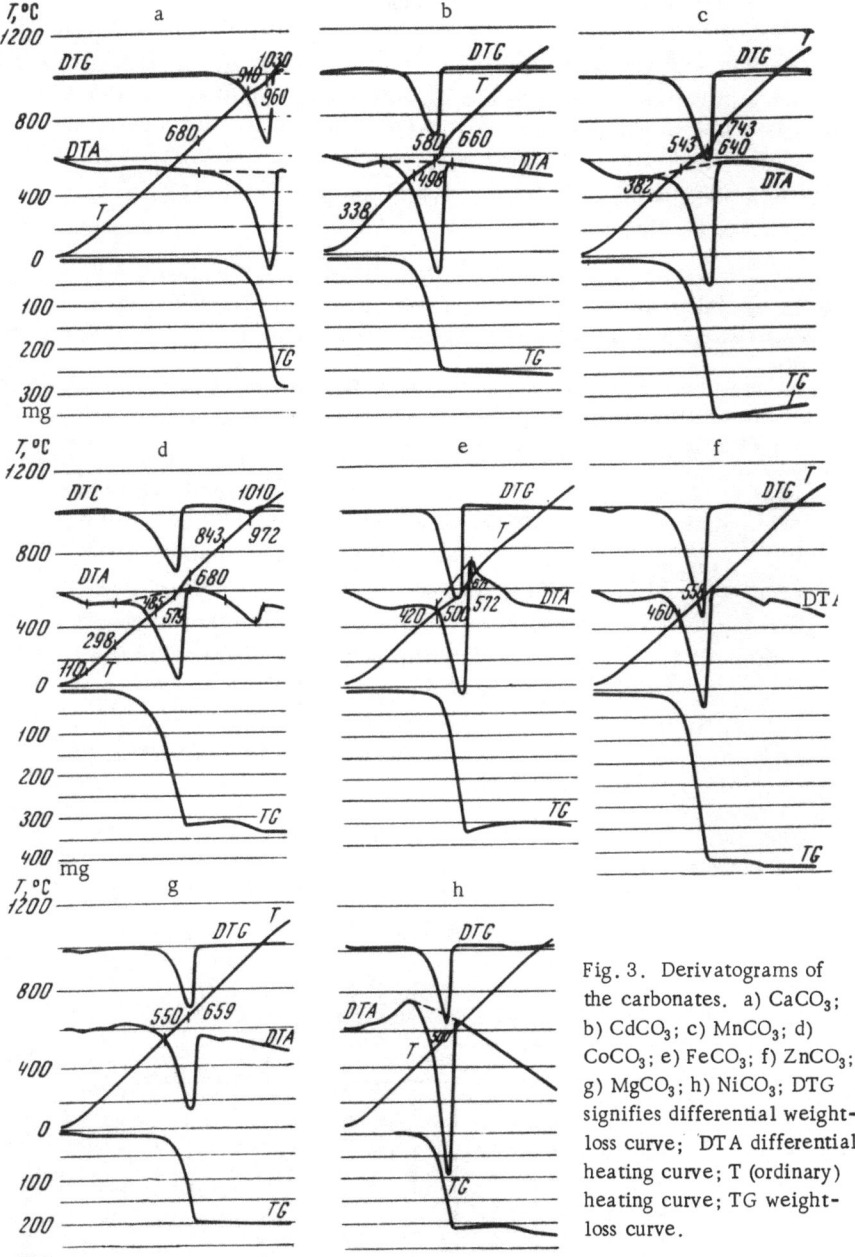

Fig. 3. Derivatograms of the carbonates. a) $CaCO_3$; b) $CdCO_3$; c) $MnCO_3$; d) $CoCO_3$; e) $FeCO_3$; f) $ZnCO_3$; g) $MgCO_3$; h) $NiCO_3$; DTG signifies differential weight-loss curve; DTA differential heating curve; T (ordinary) heating curve; TG weight-loss curve.

of the carbonate differ by no more than 1-2° for samples with
the same degree of refinement. The decomposition temperatures
of samples of the same material with different grain sizes usually
differ appreciably, mostly in the case of compounds with a high
density and structural strength. Thus in cobalt carbonate the dissociation temperature changes markedly with grain size. This
effect appears least in calcium carbonate, which does not have a
particularly strong structure.

Table 3 compares the dissociation temperatures obtained
with artificial crystals having different grain sizes (our own data)
and also the $\Delta F°$ factors of the dissociation reaction taken from [4].

The derivatograms of $CaCO_3$, $CdCO_3$, $MnCO_3$, $FeCO_3$, $ZnCO_3$,
$MgCO_3$, and $NiCO_3$ (Fig. 3) each have a single minimum corresponding to the endothermic effect. In each diagram the position of the
minimum on the DTA and DTG curves coincides; the weight loss
recorded on the TG curve equals the change in weight arising from
dissociation calculated from the molecular relationships. On the
DTA curves plotted for Mn, Fe, and Co carbonates with a high
galvanometer sensitivity (DTA = 1/200) we find an exothermic
maximum following the loss of weight on dissociation. This maximum
is associated with oxidation and the formation of the oxides of the
tervalent metals, as indicated by control Debye photographs.
On further heating the sample, the metal should be reduced to the
divalent state and the "-ous" oxide should be formed. However,

TABLE 4. Results of an X-Ray Study of the Dissociation
Products of $CoCO_3$ (URS-70)

Co_3O_4				CoO			
Reaction product after the first endothermic effect (T = 485°C)		Handbook data		Reaction product after the second endother-mic effect (T = 843°C)		Handbook data	
d, A	I/I_0, %	d, A	I/I_0, %	d, A	I/I_0, %	d, A	I/I_0, %
2.865	40	2.860	40	2.45	67	2.45	67
2.434	100	2.438	100	2.12	100	2.12	100
2.018	25	2.021	25	1.51	100	1.50	100
1.575	35	1.556	35	—	—	—	—
1.423	45	1.429	45	—	—	—	—

TABLE 5. Experimental Dissociation Temperatures and
Thermodynamic Quantities Obtained Thermochemically

Formula	T_1, °C	T_1, °K	Q, kcal	ΔH_P, kcal/mole	$\Delta S_P = \dfrac{H_P}{T_1}$
$CaCO_3$	910	1183	—40.3	40.3	34
$CdCO_3$	498	771	—34.3	34.3	44
$MnCO_3$	543	816	—40.5	40.5	49
$CoCO_3$	485	758	—35.1	35.1	46
$FeCO_3$	500	773	—28.7	28.7	37
$ZnCO_3$	460	733	—29.4	29.4	40
$MgCC_3$	550	823	—18.7	18.7	23
$NiCO_3$	500	773	—29.8	29.8	39

in the iron and manganese oxides this transformation takes place
at temperatures much higher than 1200°C. In spherocobaltite the
reduction of the oxide occurs at 843°C, and the derivatogram of
cobalt carbonate therefore differs from the others in having two
endothermic effects (Fig. 3d). One of these is associated with
the loss of CO_2, corresponding to a sample weight loss of 31.3%;
the other endothermic effect is due to the change in weight which
occurs when the tervalent cobalt oxide is transformed into the
divalent form (Table 4). The areas of the endothermic effects
formed by the DTA curves were calculated from the weights of
cut-out diagrams, and the heats of dissociation were calculated
from the Berg formula:

$$\frac{Q_1}{Q_2} = x\frac{S_1}{S_2} ,$$

where Q_1 and S_1 are the heat of dissociation and the area of the
endothermic effect of $CaCO_3$ (standard), Q_2 and S_2 are the heat
of dissociation and the area of the endothermic effect of the
carbonate under test, $x = 1.0262 + 0.000475\,\Delta T$, ΔT being a temperature difference equal to the difference between the dissociation
temperatures of the (standard) calcite and the carbonate under
consideration. The experimental values of $-Q = \Delta H$, T_1 (dissociation temperature at P = 1 atm) and also ΔS obtained by calculation are given in Table 5.

The heating curves of natural carbonates such as calcite,
magnesite, and various siderites and rhodochrosites obtained in

a Kurnakov pyrometer were given in [5]. Our own results agree
with these data. No comparison can be made with the results of
[6], since although these gave the general shape of the curves of
a number of carbonates the actual temperatures of the endothermic
effects were omitted.

The temperatures and heats of dissociation obtained from
the derivatograms were used for calculating the PT dissociation
curves of the carbonates.* The curves were calculated by the
Clapeyron-Clausius equation. The following considerations were
taken into account.

For monovariant reactions involving a gas phase, such as
reactions of decarbonatization (with phases of constant composition),
we must allow for considerable change in the volume and entropy
of the gas phase with varying T and P, i.e., the ratio $\Delta S/\Delta V$ will
be inconstant. In the calculations we consider that the volume increment in these reactions equals the volume of the gas phase,
while the change in the volume of the solid phases may be neglected.
We further consider that the equation of state for ideal gases may
be applied to the gas phase and that the ΔH of the reaction is constant.

The Clapeyron–Clausius equation is applicable to all the
monovariant curves; it may be derived from the differential equation
of the isobaric-isothermal potential (Gibbs free energy) $dZ =
-SdT + VdP$ by the method of displaced equilibrium [8].

Remembering that for every reversible monovariant phase
transformation the potential Z must necessarily be the same in
the initial and final state ($Z_1 = Z_2$), the same applying to the in-
crements $dZ_1 = dZ_2$ when the external conditions are varied, we find

$$dZ_2 - dZ_1 = -(S_2 - S_1)\,dT + (V_2 - V_1)\,dP = 0.$$

If we denote $S_2-S_1 = \Delta S$ and $V_2-V_1 = \Delta V$, we obtain $\Delta SdT = \Delta VdP$
(the Clapeyron–Clausius equation). Since for an ideal gas $\Delta V \cdot P =
RT$, we have

$$\Delta SdT = \frac{RT}{P}\,dP,$$

or

$$\frac{dP}{P} = \frac{\Delta S}{RT}\,dT.$$

---

*The dissociation reactions of several carbonates were calculated in [7] using the
approximate Nernst formula.

And since $dP/P = d \ln P$, we have $d \ln P = (\Delta S/RT)\, dT$.

When equilibrium has been established for the dissociation reaction

$$\Delta S = \frac{\Delta H}{T},$$

and then

$$d \ln P = \frac{\Delta H}{RT^2}\, dT = \frac{\Delta H}{R}\, T^{-2} dT.$$

Integrating this equation from $P_1 = 1$ atm to the unknown pressure P and from T_1 (corresponding to the pressure P_1) to the temperature T, and treating $\Delta H = $ const, we obtain

$$\int_1^P d \ln P = \frac{\Delta H}{R} \int_{T_1}^T T^{-2} dT,$$

$$\ln P = \frac{\Delta H}{R}\left(\frac{1}{T_1} - \frac{1}{T}\right) = \frac{\Delta H (T - T_1)}{RTT_1},$$

or, substituting for the value of R and converting to base 10 logarithms,

$$\log P = \frac{\Delta H (T - T_1)}{4.576 T_1 T},$$

TABLE 6. Parameters of the PT Dissociation Curves of Artificial Carbonates

CaCO₃		CdCO₃		MnCO₃		CoCO₃		FeCO₃		MgCO₃		ZnCO₃		NiCO₃	
T, °C	P, atm	T, °C	P, atm	T, °C	P, atm	T, °C	P, atm	T, °C	P, atm	T, °C	P, atm	T, °C	P, atm	T, °C	P, atm
910	1	498	1	543	1	485	1	500	1	550	1	460	1	500	1
960	2.016	550	4.113	593	5.12	520	1.943	550	3.11	600	1.93	500	2.84	550	2.13
1000	3.394	600	13.68	625	9.781	550	6.296	660	7.06	650	3.46	560	11.27	600	9.1
1050	6.227	650	39.90	650	18.07	570	9.9	650	20.8	700	5.84	600	25.47	650	22.9
1100	10.94	670	59.29	675	32.36	600	21.48	700	46.45	800	14.39	620	37.15	700	53.0
1150	18.45	700	104.2	700	56.23	620	33.88	750	85.21	900	30.48	660	75.51	750	122.0
1200	30.06	725	185.2	750	146.6	630	42.17	800	181.2	950	42.36	700	144.9	800	226.8
1242	44.11	750	248.3	770	229.6	670	96.83			1000	57.28	760	349.9		
1300	72.78					700	172.2			1050	75.68				
1350	108.4									1100	97.50				
1400	158.1									1200	156.7				
1450	235.4									1250	183.3				
										1300	235				

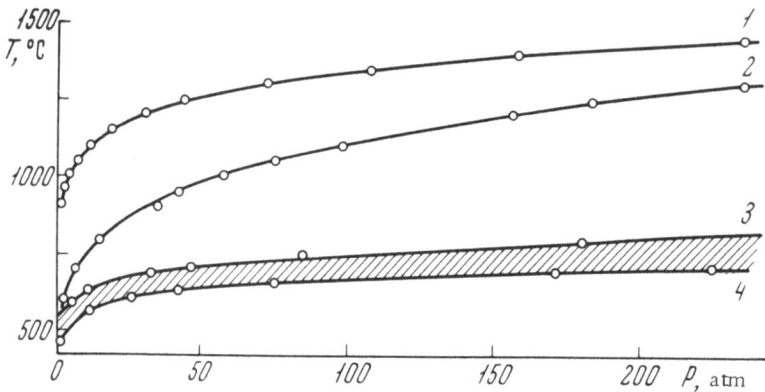

Fig. 4. Calculated dissociation curves of trigonal carbonates. 1) $CaCO_3$; 2) $MgCO_3$; 3,
4) region corresponding to the carbonates of the transition elements.

where R ln x = 4.576. This equation may be used for approximately
calculating the equilibrium curves in decarbonatization reactions
from the experimental data listed in Table 5. The coordinates of
the points on the dissociation curves of the carbonates calculated   –
from Eq. (1) are given in Table 6. Figure 4 shows the dissociation
curves of calcite and magnesite, and also indicates the region
corresponding to the carbonates of the transition elements. The
position of each curve is shown on a larger scale in Fig. 5.

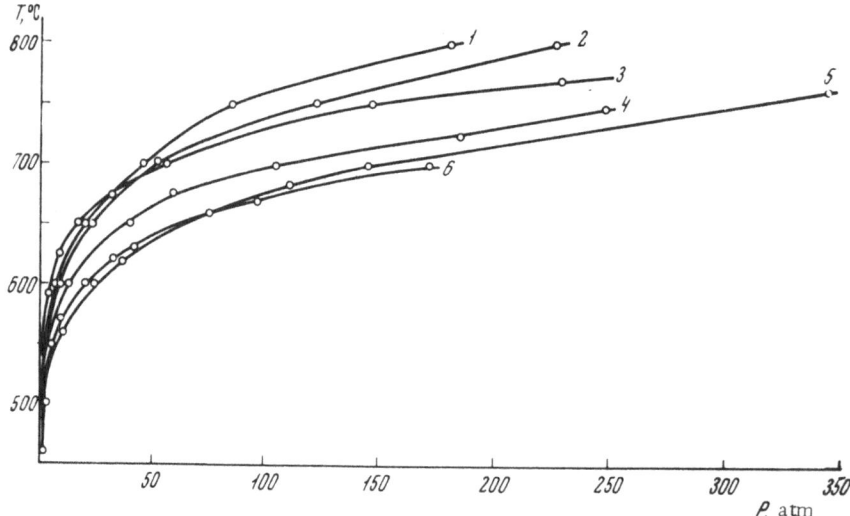

Fig. 5. Calculated dissociation curves of the carbonates of the transition elements. 1)
$FeCO_3$; 2) $NiCO_3$; 3) $MnCO_3$; 4) $CdCO_3$; 5) $ZnCO_3$; 6) $CoCO_3$.

The calculated dissociation curves of the carbonates may be used for choosing the parameters to be used in growing crystals from the melt in a CO_2 atmosphere.

Literature Cited

1.   P. J. Wyllie, J. Amer. Ceram. Soc., 50:43 (1967).
2.   E. H. Baker, J. Chem. Soc., 2:464 (1962).
3.   F. H. Smith and Z. H. Adams, J. Am. Chem. Soc., 45:1167 (1923).
4.   W. Latimer, Oxidation States of the Elements and Their Potentials in Aqueous Solutions [Russian translation], IL (1954).
5.   L. G. Berg and A. V. Nikolaev, Thermography [in Russian], Izd. AN SSSR (1945), pp. 88.
6.   V. P. Ivanova, Zap. Min. Obshch., Ser., II, Vol. 90, No. 1 (1961).
7.   D. S. Korzhinskii, Zap. Min. Obshch., Ser. II, Vol. 64 (1935).
8.   I. Prigozhin and R. Defei, Chemical Thermodynamics [in Russian], Izd. Nauka Novosibirsk, (1966), p. 261.

Solubility of Sodalite in Aqueous Solutions of NaOH under Hydrothermal Conditions

L. N. Dem'yanets, E. N. Emel'yanova, and O. K. Mel'nikov

Introduction

The mineral known as sodalite $Na_8 [Al_6Si_6O_{24}] Cl_2$ is crystal-chemically one of the group of skeletal aluminosilicates with zeolitic properties. Recently sodalite single crystals have attracted the attention of physicists in connection with the interesting optical properties of these crystals [1]. We ourselves have grown sodalite crystals by the hydrothermal method, which is the most promising method for the production of single crystals, since any attempt at growing sodalite from the melt simply produces glass of an analogous composition, while growing from solution in the melt only yields small crystals (under 2 mm) [2].

The production of large single crystals under hydrothermal conditions is very long and cumbersome. Some experiments last six months. Such experiments have therefore to be carried out under conditions in which the main factor determining the growth of the crystal, the supersaturation ΔS in the growth zone, is exactly known, and so are the changes taking place in ΔS during the experiment (i.e., changes which occur while the working conditions are being established, during any fluctuations in working conditions, and as the operation is being brought to a close).

The practical conditions of the hydrothermal process are not such as will permit direct observation of the dissolution and growth of the crystal. Analysis of the process usually starts with a study of the finished crystal, and various hypotheses as to the manner

in which the process takes place are derived from this. A study
of these crystals provides data as to the rate of crystallization
in various directions and under various conditions, as well as other
data relating to mass transfer in the autoclave and the quantitative
relationship between the phases. By analyzing the solution after
the experiment we may determine the quantities characterizing
the "residual" (usually metastable) concentration of dissolved
material. However, the first stage in the growth of the crystal
(often determining the whole course of the reaction), namely, the
dissolution of the charge, remains quite unkown; so does the quantitative supersaturation in the growth zone, the dependence of the
solubility on various factors, and so forth. All these quantities
may be determined by studying the solubility of the crystallizing
compound in relation to the composition of the solvent, the temperature, and the pressure.

It is quite a difficult matter to study the solubility of crystals under hydrothermal conditions if the compound in question
does not form any metastable solutions over a wide temperature
range. If there is a considerable metastable range of this kind,
however, we may use the least troublesome method of studying
solubility at high temperatures and pressures − the method of weight
loss. In this case quenching ensures the maintenance of conditions
in the autoclave corresponding to equilibrium conditions at the
test temperature.

When studying the growth and solubility of silicates and
aluminosilicates, additional difficulties arise; these were first
noticed when studying the solubility of quartz. The difficulties
principally arise from the phase separation of the solution under
certain conditions, and the formation of a "heavy" phase enriched
with silicon [3, 4]. This may possibly explain the small number
of detailed investigations which have been made into the solubility
of silicates and aluminosilicates under hydrothermal conditions.
Thus, for example, it was shown in [5], in connection with the
solubility of albite in H_2O at 200-350°C, that under these conditions
there was a partial decomposition of the albite, with the formation
of analcime; hence in this case the true solubility of albite in water
was not really being determined. We studied the solubility of
sodalite under conditions similar to its conditions of synthesis, so
as to determine the optimum region for the crystallization of
sodalite on a seed.

Experimental Method and Apparatus

The solubility of sodalite was determined by the weight-loss method [5] using single-crystal blocks in contact with aqueous solutions of NaOH of a specified concentration.

In order to determine the weight loss of the seed we used autoclaves of about 40 cm^3 in volume (Fig. 1) made of 45KhNMFA steel. The autoclaves have a standard conical closure of the self-sealing type; the obturator is furnished with a thermometric "pocket" for measuring the temperature in the middle of the control autoclave.

The experiments were carried out in very corrosive media (aqueous solutions of NaOH up to 40 wt.% concentration), and we therefore had to use protective linings. These were made of silver or Teflon (Fig. 2), being hermetized in the first case by means of silvered copper plates 0.05 mm thick and in the second by means of Teflon stoppers. The seeds were fixed in a special frame made of sheet silver or directly in the top of the lining, so that the seed

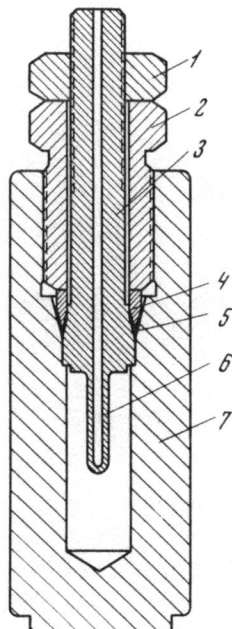

Fig. 1. Autoclave for studying solubility by the quenching method. 1) Stop nut; 2) pressure nut; 3) obturator; 4) steel ring; 5) copper gasket; 6) thermometric "pocket"; 7) body of autoclave.

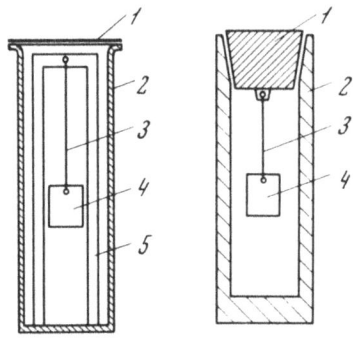

Fig. 2. Silver (a) and Teflon (b) lining-type inner vessels. 1) Cover; 2) body of lining; 3) silver wire; 4) seed block; 5) silver frame.

crystals lay in the middle of the latter. The autoclaves were filled with a specified quantity of solution and placed in a furnace with an electric heater (Fig. 3).

The furnace is furnished with an automatically opening cover plate situated underneath so as to enable the furnace to be rapidly discharged in order to quench the autoclave in running water.

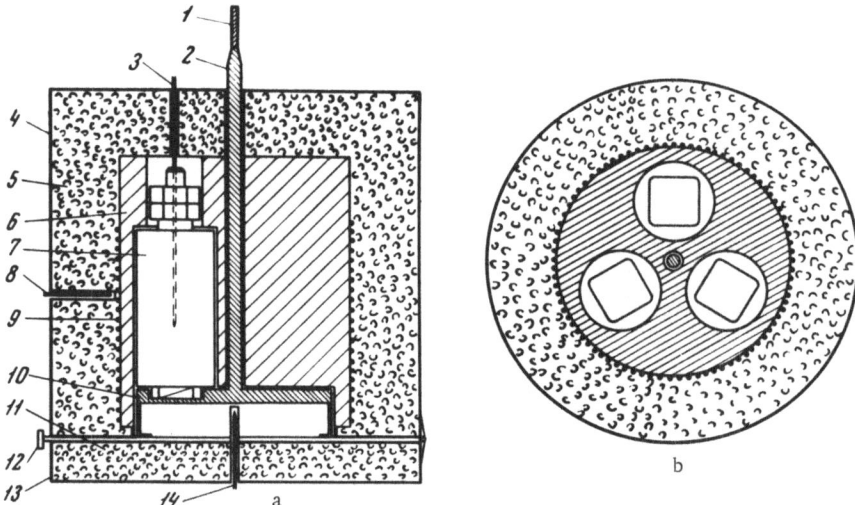

Fig. 3. Furnace for studying solubility by the quenching method. a) Side view; b) plan; 1) cable; 2) carriage holder; 3) lead for measuring thermocouple; 4) body of furnace; 5) thermal insulation (asbestos dust); 6) cast iron; 7) autoclave; 8) lead for regulating thermocouple; 9) heater; 10) autoclave carriage; 11) sheet asbestos; 12) lock; 13) flap cover; 14) lead for recording thermocouple.

The furnace accommodates three autoclaves, one of which has a control thermocouple. In order to provide more uniform heating the autoclaves are sited inside a cast iron block.

The temperature was measured at two points: inside the control autoclave and in the lower part of the autoclave carriage (Fig. 3). The temperature was measured with a chromel–alumel thermocouple and recorded with a KP-64 recorder. The temperature was regulated with an electronic regulator of the ÉRM type: the regulating thermocouple was placed close to the heater.

Preliminary measurement of the temperatures along the autoclave and at the top and bottom of the furnace showed that there was no appreciable temperature gradient; in the majority of the experiments the temperature in the lower part of the autoclave carriage was therefore not recorded. Temperature fluctuations during the experiment amounted to $\pm 2°C$. The time required to bring the furnace to the steady state lay between five and twelve hours, depending on the temperature specified.

Usually the specified temperature was approached "from below." In some experiments a superheating of 8–10° was first applied and the temperature was then reduced to the specified value and held for an appropriate period. Under these conditions characteristic signs of the growth of faces were noted on the surface of the crystal. The results obtained in the two cases were identical, i.e., the whole excess of sodalite passing into solution on superheating crystallized on the seed after subsequent cooling.

Some of the experiments were carried out at saturated vapor pressures (at 200-250°C), in others the vapor phase was absent and the pressure was calculated from the autoclave occupation factor.

In order to determine the solubility of sodalite we used seeds made from artificial hydrosodalite crystals. In the first stages of the work we used natural sodalite or blocks of artificial sodalite with an interlayer of natural seed material. Under natural conditions only ordinary chlorosodalite $Na_8[Al_6Si_6O_{24}]Cl_2$ or a sulfur-containing variety, hackmanite $Na_8[Al_6Si_6O_{24}](Cl, S)_2$, form large single-crystal blocks. On synthesis under hydrothermal conditions, in the absence of Cl^- ions hydrosodalite $Na_8[Al_6Si_6O_{24}](OH)_2 \cdot nH_2O$ is formed, where n = 1-3. Hydrosodalite is also encountered in Nature, but only in the form of fine grains. Hence, in studying solubility, either natural sodalite, usually containing various impurities, may be used as original material, or else artificial hydrosodalite. In order to study solubility the original hydrosodalite was obtained by crystallization in a 30% aqueous solution of NaOH with a temperature of the order of 250°C in the growth zone and a pressure of $\sim$400 atm.

From the artificial sodalite we cut blocks $5 \times 10 \times 15$ to $10 \times 10 \times 20$ mm in size parallel to the faces of the cube (100), rhombododecahedron (110), tetrahedron (111), and trigontritetrahedron (211). The solubility of natural sodalite was studied with nonoriented samples owing to the difficulties of orientation. The resultant blocks were washed with ethyl alcohol and brought to constant weight at 80°C. It should be noted that the original artificial material contained nonuniformly developed cracks, and the natural sodalite contained a considerable amount of impurities, which might affect the results of the solubility determinations.

The aqueous solutions were prepared from chemically pure NaOH, and the specified concentration was achieved by diluting the saturated solution with water. In these solutions we determined the density at 15°C and the Na_2O content. The NaOH solutions were stored in polyethylene vessels so as not to be contaminated with SiO_2.

After the quenching of the autoclave, samples were taken in certain cases so as to determine the Na, Si, and Al content. The samples were taken from differnt heights of the autoclave so as to determine the average solubility, in case of any possible phase separation of the solution. The sodalite seeds were washed in ethyl alcohol* and dried to constant weight at 80°C. The crystal weight loss was referred to the volume of solvent taken $S = (P_2 - P_1) \cdot 100 / V_1$, where P_2 is the weight of the seed block before the experiment, P_1 is the weight after the experiment, V_1 is the original volume of the NaOH solution. We determined the solubility in wt.% $S = (P_2 - P_1) \cdot 100 / (P_2 + P_0)$ (P_0 is the original weight of the mineralizer) and molar % of the dissolved substance $S = M_1 / (M_1 + M_2 + M_3)$, where M_1 is the number of moles of dissolved sodalite in

---

*On washing with water, sodium was leached out owing to the zeolitic character of the sodalite structure.

100 g of solution, M_2 is the number of moles of NaOH, and M_3 is the number of moles of H_2O.

In a number of cases the amount of dissolved material was determined experimentally by finding the relationship $S' = f(t)_{C = const}$ (S' is the solubility under nonequilibrium conditions and t is the time) for a constant occupation factor, temperature, solvent concentration, and area of the seed blocks. The time between the opening of the furnace and quenching was 5 sec; cooling the autoclave to room temperature in running water took 5 min.

Results of the Experiments

We studied the solubility of sodalite in aqueous solutions of NaOH (10-40 wt.%) between 200 and 300°C. The results appear in Table 1.

The occupation factor was varied between 0.55 and 0.90 during the experiments. Preliminary investigations showed that, in the temperature and concentration ranges studied, the solubility of sodalite was almost independent of pressure.

Using data taken from [7, 8] regarding the densities and pressure of aqueous solutions of NaOH at high temperatures, and data taken from [9] regarding the specific volumes of H_2O, we may determine the relation between the vapor and liquid phases at various temperatures for various autoclave occupation factors (Fig. 4). For 90% occupation the vapor phase is practically absent even at 200°C for a 30% solution of NaOH. The homogenization temperature increases with falling occupation factor. In the majority of cases the experiments were carried out in the absence of the vapor phase. In those experiments in which the vapor phase was present, a correction was introduced for the water passing into the vapor under the conditions of the experiment. At 200-300°C only H_2O passed appreciably into the vapor phase, the maximum amount in the vapor being 0.04 g. The correction for the NaOH concentration was no greater than 0.1% for concentrations of the order of 30%. Table 1 gives the NaOH concentrations at the temperature of the experiments allowing for the vapor-phase correction.

The experimentally-determined period during which the autoclaves had to be held under the specified conditions in order to reach equilibrium was some 4-5 days for a 30% solution of NaOH

TABLE 1, Results of an Investigation Into the Solubility of
Sodalite under Hydrothermal Conditions

Serial No.	T_{exp},°C	Occupation factor f, %	Concentration of NaOH at T_{exp}		Solubility of sodalite		
			wt.%	molarity	g/100 ml NaOH	wt.%	mol.%
1	205	83.8	20.2	6.25	1.22	0.99	0.0209
2	203	87.4	20.2	6.25	1.32	1.01	0.0213
3	203	83.0	20.2	6.25	1.40	1.13	0.0239
4 *	202	56.6	20.23	6.25	1.13	0.91	0.0192
5	200	93.5	20.2	6.25	1.30	1.05	0.0222
6	199	83.0	10.2	2.37	0.62	0.55	0.0110
7 *	199	53.8	10.33	2.87	0.53	0.52	0.0103
8	202	87.0	10.3	2.87	0.73	0.65	0.0129
9	198	85.9	37.65	15.10	2.14	1.50	0.0357
10 *	198	55.0	37.7	15.10	2.23	1.53	0.0371
11	202	84.7	37.65	15.10	2.37	1.58	0.0376
12 *	202	62.2	37.7	15.10	2.15	1.52	0.0362
13	200	90.0	29.96	10.69	1.66	1.24	0.0282
14	200	90.0	29.96	10.69	1.69	1.26	0.0234
15	200	90.0	29.96	10.69	1.74	1.29	0.0290
16	200	90.0	19.65	6.43	1.08	0.85	0.0179
17	200	90.0	19.65	6.43	1.12	0.91	0.0192
18	200	90.0	19.65	6.43	1.13	0.91	0.0192
19	250	90.0	29.96	10.69	2.65	1.95	0.0465
20	250	90.0	19.65	6.43	1.56	1.25	0.0284
21	250	90.0	19.65	6.43	1.59	1.29	0.0272
22	300	90.2	30.20	10.81	4.45	3.24	0.0746
23	300	79.7	30.20	10.81	4.06	2.97	0.0630
24	300	80.0	25.5	8.60	2.84	2.15	0.0474
25	300	80.0	25.50	8.60	2.90	2.21	0.0483
26	300	80.0	25.50	8.60	2.67	2.04	0.0450
27	300	80.1	20.15	6.31	2.05	1.65	0.0350
23 *	285	78.5	20.15	6.31	1.96	1.55	0.0329
29	300	81.7	20.15	6.31	2.04	1.64	0.0347
30 *	300	77.9	17.0	5.12	1.80	1.45	0.0300
31	285	78.9	17.0	5.12	1.47	1.24	0.0257
32 *	300	68.0	15.10	4.48	1.47	1.24	0.0257
33 *	300	77.0	15.10	4.48	1.47	1.23	0.0252
34	300	79.8	15.10	4.48	1.47	1.23	0.0252
35 *	300	75.6	15.10	4.48	1.50	1.27	0.0260
36	300	78.4	15.10	4.48	1.46	1.23	0.0252
37	300	80.0	15.10	4.48	1.56	1.31	0.0269
38 *	300	71.4	10.30	2.87	1.23	1.08	0.0215
39 *	300	69.8	10.30	2.87	1.17	1.03	0.0204
40	300	81.0	H_2O	—	0.84	0.84	0.0156
41	200	87.2	30.20	10.81	1.68	1.24	0.0280
42 *	198	67.3	30.20	10.81	1.67	1.24	0.0280
43	210	91.9	30.20	10.81	1.81	1.42	0.0320
44 *	205	62.9	30.20	10.81	1.74	1.28	0.0289
45	210	86.4	30.20	10.81	1.90	1.44	0.0325

*Experiments in the presence of the vapor phase.

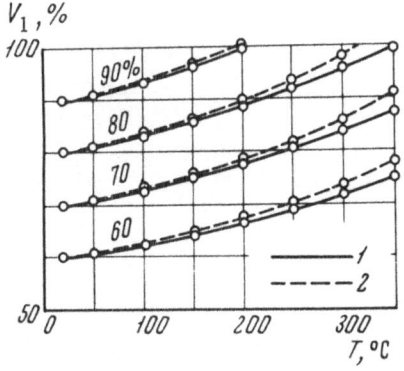

Fig. 4. Relation between the liquid and vapor phases at 25-350°C. 1) 30% NaOH; 2) 20% NaOH; figures on the curves give the occupation factors at room temperature; V_1 is the percentage volume occupied by the liquid phase.

at 200-250°C (Fig. 5). The long duration of the experiment is due to the absence of forced mixing and the great viscosity of the medium. At higher temperatures equilibrium is achieved much more rapidly. All the experiments were, as a rule, carried out with holding periods of about six days.

In studying the solubility we used concentrated solutions of NaOH, since the field of stability of sodalite depended substantially on the NaOH concentration. In the $Na_2O-Al_2O_3-SiO_2-H_2O$ system, increasing the Na_2O concentration leads to a regular change in the stable phases; these appear in the order nepheline, cancrinite, sodalite. The range of stable crystallization of sodalite lies at NaOH concentrations of 20 wt.% and over. The lower limit of the formation of sodalite varies with the temperature of the experiment and the composition of the seed crystals. The high viscosity of of 20-40% aqueous solutions of NaOH and the absence of agitation partly explain the fact that the upper part of the seed blocks dissolve more vigorously. Concentration flows evidently prevent the

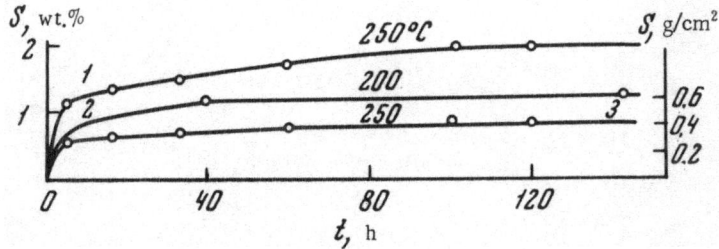

Fig. 5. Curves representing the experimentally determined time required to establish equilibrium for 30% NaOH. 1), 2) Seed weight losses in wt.% (left-hand ordinate axis); 3) seed weight losses referred to unit surface (right-hand ordinate axis).

intensive dissolution of the lower part of the seed (F:~ 6). It was found in a number of experiments that the solution contained different proportions of Na, Si, and Al at different heights in the autoclave (in experiments involving a holding period of four days in the working mode). The difference in the Na_2O concentration in the upper and lower parts of the autoclave reached a maximum of 0.97% for an NaOH concentration of 30 wt.%.

Different SiO_2 concentrations were also observed when studying the solubility of silica in H_2O [10]: In the lower zone of the autoclave three or four times as much dissolved SiO_2 was found (in the lower part ~0.12 g SiO_2 were determined in 100 parts of H_2O at 300°C and in the upper part ~0.03 g SiO_2). Roughly the same difference in dissolved material concentration was found in different zones of the autoclave when studying the solubility of albite in H_2O at 200–300°C and at pressures up to 300 bar [5]. In the latter case the albite decomposed with the formation of a zeolitic phase, so that the concept of albite solubility was somewhat arbitrary.

In the case of sodalite the difference in the concentrations of the solution obtained on analyzing samples taken from the upper and lower zones of the autoclave was much smaller. On referring the analysis to the amount of sodalite in 100 ml of solvent, the maximum difference is ~0.4 g for a total concentration of dissolved sodalite equal to 4 g.

On performing an analysis of the solution obtained after quenching autoclaves held in the working mode for longer periods

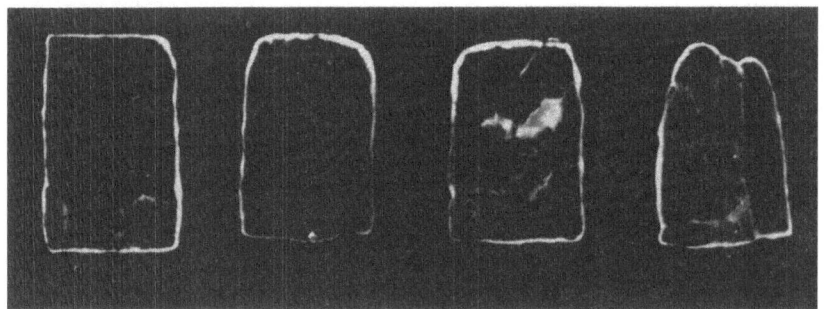

Fig. 6. Dissolution of sodalite seed blocks. Times of holding in the working mode 2, 16, 29, and 65 h respectively.

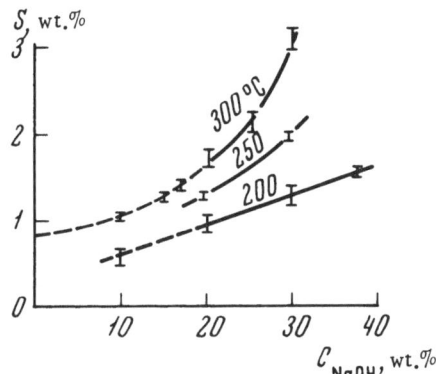

Fig. 7. Relation between the solubility of sodalite and the NaOH concentration at constant temperature.

(5-6 days) samples taken from different levels in the autoclave gave identical solubilities.

Apparently on increasing the period of the experiment the concentrations of the dissolved substance at different levels of the autoclave become equalized as a result of diffusion. By way of example we may give the results of an analysis of the sample in experiment No. 23 (Table 1). The solubility obtained was 4.06 g per 100 ml of solvent; an analysis of a sample from the upper part of the autoclave gave 4.04 g and that of a sample from the lower part of the autoclave gave 4.02 g of sodalite. The spread in these values lies within the limits of experimental error.

In our experiments we observed no formation of a "heavy" phase, such as is characteristic of the dissolution of SiO_2 under hydrothermal conditions. The solubility of sodalite increases with increasing concentration of the solvent at constant temperatures (Fig. 7). The slope of the curves relating solubility to NaOH concentration depends on the temperature: The higher the temperature, the more sharply does the curve rise.

The solubility curve of natural sodalite (hackmanite) lies rather below the curve plotted for the solubility of artificial sodalite. The difference between the solubilities of hydrosodalite and hackmanite is of the order of the experimental error, but the hackmanite solubility always appears ~0.1 wt.% below the solubility of hydrosodalite. A greater difference appears in the rates of dissolution; the artificial mineral dissolves considerably more rapidly.

At 200°C the dependence of the solubility of sodalite (S) on the NaOH concentration is of a linear nature $S = \alpha C_{NaOH} + S_1$, where S_1

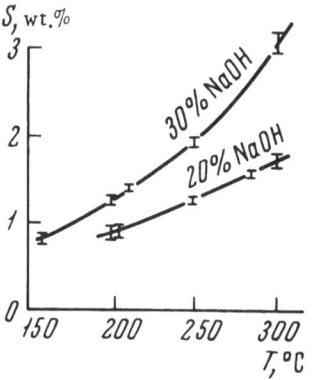

Fig. 8. Temperature dependence of the solubility of sodalite for constant NaOH concentration.

is the solubility of sodalite in water at T = 200°C, α is a coefficient equal to 0.034; S_1, determined by extrapolating the straight line to the S axis, equals ∼0.258. At 300°C the dependence of the solubility of sodalite on the NaOH concentration is parabolic. In Fig. 7 the broken part of the curves in the concentration range 0–20 wt.% NaOH characterizes the nominal solubility of sodalite, since in this range of concentrations sodalite does not constitute a stable solid phase. In a 10–15% solution of NaOH, the sodalite seed blocks are covered with a film of hydrocancrinite. The data relating to this range of concentrations characterize the average solubility of aluminosilicates with a ratio $SiO_2 : Al_2O_3 = 2 : 1$ (the composition of hydrosodalite is $Na_8[Al_6Si_6O_{24}] \ (OH)_2 \cdot nH_2O$, of hydrocancrinite $Na_8[Al_6Si_6O_{24}] \ (OH)_2 \cdot nH_2O$, and of nepheline $Na[AlSiO_4]$).

For a constant concentration of the solvent, an increase in temperature causes a rise in the sodalite solubility (Fig. 8). In the concentration and temperature range studied the temperature coefficient of the solubility is positive. The higher the NaOH concentration, the steeper is the $S = f(T)_{P = \text{const}}$ curve, and the greater is the deviation from linearity. The absolute increment in solubility on changing the temperature by 100°C ($T_1 = 200°C$, $T_2 = 300°C$) in 20% NaOH is ∼0.8 wt.% and in 30% NaOH ∼1.8 wt.%. We shall subsequently discuss the solubility of sodalite in a 30% NaOH solution, since this concentration characterizes the field of stability of sodalite.

A consequence of the nonlinear temperature dependence of the solubility in a 30% NaOH solution is a nonuniform change in the supersaturation ΔS with temperature for a constant temperature

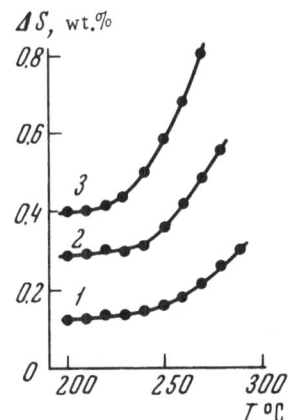

Fig. 9. Temperature dependence of the supersaturation for a constant temperature drop and 30% NaOH. 1) $\Delta T = 10°$; 2) $\Delta T = 20°$; 3) $\Delta T = 30°$.

drop ΔT. The supersaturation ΔS is calculated as the difference in the solubilities of sodalite at the two temperatures determining ΔT (Fig. 9). The horizontal axis gives the temperatures of the growth zone $T_1 (T_1 < T_2)$. The form of the curves for $\Delta T = 10, 20, 30°$ is practically identical; the $\Delta S = f(T)_{\Delta T = const}$ relationship is almost linear, and the slope of the curves increases with increasing ΔT. As temperature falls the curves relating the supersaturation to temperature for a constant temperature drop flatten out.

Figure 10 illustrates the relation between ΔS and temperature drop for various T and ΔT. In order to obtain the same supersaturation at different temperatures a considerably smaller drop is required for $T = 270$ than for $T = 200°C$. A supersaturation of $\Delta S = 0.15$ wt.% of sodalite may be obtained at $200°C$ for $\Delta T = 12°$ or at $270°C$ for a ΔT of the order of $6°$; a supersaturation of 0.4 wt.% occurs in a 30% solution of NaOH at $T = 220°C$ and $\Delta T = 27°$, or at $T = 270°C$ and $\Delta T = 16°$.

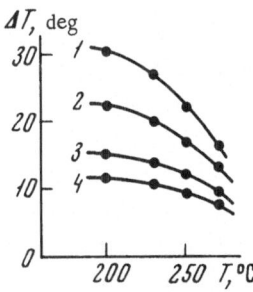

Fig. 10. Curves of equal supersaturation for variable T and ΔT and 30% NaOH. Supersaturation ΔS, wt.%: 1) 0.4; 2) 0.3; 3) 0.2; 4) 0.15.

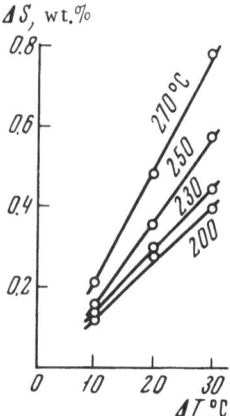

Fig. 11. Dependence of the supersaturation (ΔS) on the temperature drop at constant temperature in 30% NaOH.

In order to grow crystals it is important to know the way in
which the supersaturation varies with increasing temperature drop.
We see from Fig. 11 that the supersaturation varies almost linearly
with increasing ΔT: The higher the ΔT the more do the $\Delta S =$
$f(\Delta T)_{T = \text{const}}$ lines diverge.

It is well known that for α quartz a temperature drop of 15°
corresponds to a supersaturation of ~ 0.07 wt.% in a $0.51 M$ solution of NaOH at 300-400°C and a pressure of 1390 bar [11]. Unfortunately there are no data relating to the solubility of SiO_2 and
Al_2O_3 under the conditions used for growing sodalite (200-300°C,
400 atm, NaOH concentration of the order of 30 wt.%). A solubility
close to the solubility of sodalite at 300°C (~ 4 wt.%) occurs for
quartz in a 17% solution of NaOH at 250-300°C [10, 11]. In order
to achieve a sodalite solubility quantitatively comparable with
the solubility of quartz and alumina [10-12] far more alkaline solutions are required.

In the $Na_8[Al_6Si_6O_{24}]$ $(OH)_2 \cdot nH_2O - Na_2O - H_2O$ system the temperature dependence of the solubility of sodalite in the temperature range 200-300°C closely follows the Van't Hoff law:

$$\frac{d\ln K}{dT} = -\frac{\Delta H}{RT^2},$$

where K is the equilibrium solubility of sodalite, ΔH is the heat of
dissolution. Figure 12 shows the dependence of log S (the solubility

is expressed in mol. %) on the reciprocal temperature for various NaOH concentrations.

From the slopes of the straight lines we may calculate the heat of dissolution of sodalite; the values for 30, 20, and 10% solutions are respectively 4.9 ± 0.5; 3.4 ± 0.5 and 2.4 ± 0.5 kcal/mole.

On the basis of the values of ΔH so obtained we may calculate the entropies of dissolution expressed as $\Delta H/T°K$ under equilibrium conditions.

The application of the foregoing quantitative data regarding the solubility of sodalite to the kinetics of crystal growth on a seed will be considered subsequently.

Figure 13 shows the total consumption of the charge F and the increment in the amount of material on the seed per day M calculated from data relating to the solubility of sodalite and the growth rates of the faces: $F = M + S$, where S is the solubility of sodalite under the specified conditions.

The resultant curves are somewhat arbitrary, as we have taken the average growth rate of the rapidly-growing (111) and (100) faces for different ΔT. The values of F and M are calculated for 1 liter of solvent (30% NaOH) and a constant initial surface area of the seeds equal to 30 cm^2 (the surface area of the seeds in the experiments on growing sodalite crystals was $\sim$30-33 cm^2). The

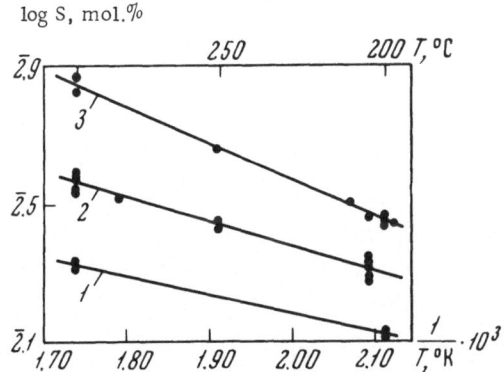

Fig. 12. Dependence of log S on reciprocal temperature. NaOH concentration (wt.%): 1) 10; 2) 20; 3) 30.

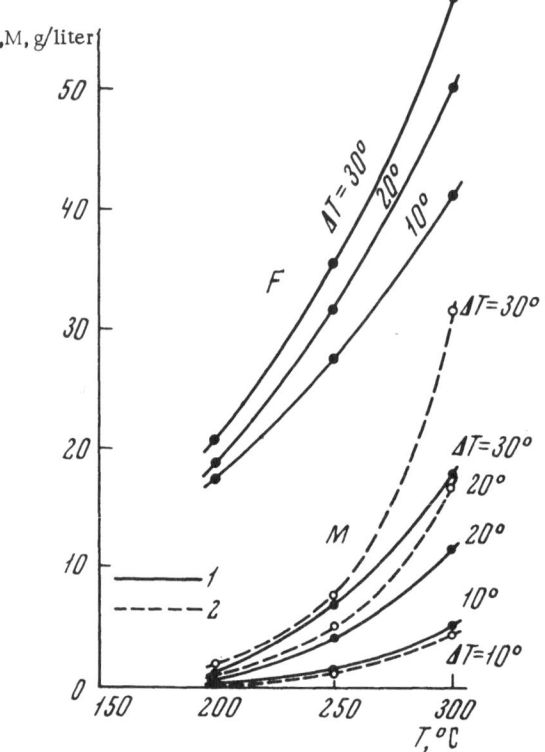

Fig. 13. Change in the consumption of the charge
F and the mass increment on the seed M with changing temperature for a constant temperature drop. 1)
Theoretical curves based on sodalite solubility data;
2) calculated data based on the growth of the sodalite
crystals.

increment of material on the seed M, expressed in g/day, is calculated as the product of the average growth rate v, the surface
area of the seeds, and the specific gravity of sodalite.

We see from Fig. 13 that on increasing the temperature in
the growth zone by 100°, the consumption of charge increases by
a factor of 2–3, depending on the value of ΔT.

In setting up long experiments on growing sodalite single
crystals, the curves so obtained enables us to estimate the amount
of charge required in order to maintain the specified supersaturation
in the growth zone.

Figure 13 gives the mass-transfer curves $M = f(T)_{\Delta T = const}$
(broken) calculated from experimental data relating to the weight
loss of the charge used for growing the sodalite crystals. The
continuous lines indicate the calculated curves for the same areas
of the seeds, temperature, pressure, and concentration of the solution. We see that for a slight supersaturation ($\Delta T = 10°$) the theoretical curve practically coincides with the experimental. For large
supersaturations the experimental curve deviates from the theoretical, owing to spontaneous crystallization, which sharply alters
the surface area of the sodalite seeds taken for the calculation.

The growth rate of the various faces of sodalite crystals increases linearly with increasing ΔT. Since in the temperature
range considered the supersaturation ΔS depends almost linearly
on the temperature drop ΔT (Fig. 11), we have $v = k\Delta S$ (Fig. 14).
The growth rates in the [111], [100], [110] directions increase
linearly with increasing supersaturation ΔS. A change in the temperature of the growth zone T_1 from 200 to 250° leads to a considerable increase in the growth rates of the (111), (100), and (110)
faces.

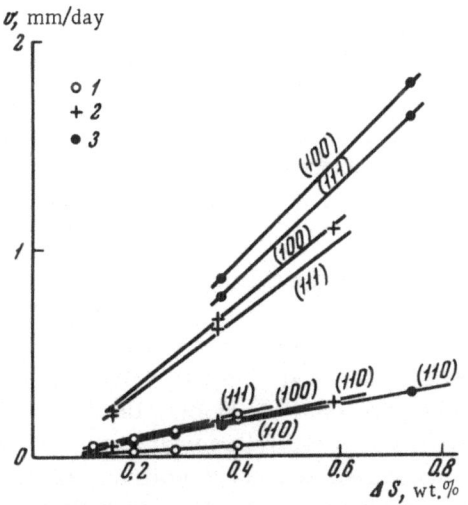

Fig. 14. Dependence of the growth rates of sodalite (111), (100), and (110) faces on the supersaturation at constant temperature. 1) Temperature of
growth zone 200°C, 2) 250°C; 3) 300°C.

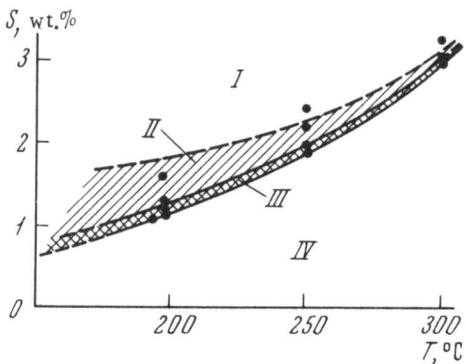

Fig. 15. Phase diagram of the solubility of sodalite under hydrothermal conditions in 30% NaOH.
I) Labile region; II) metastable region; III) region
of critical supersaturations; IV) region of unsaturated
solutions.

On comparing the growth rates of the rapidly-growing (100)
and (111) faces and the faces of the rhombododecahedron (110) we
see that, for a constant temperature and variable supersaturation,
the ratio $\bar{v}_{[111],[100]}/v_{[110]}$ remains almost constant. With increasing
temperature the difference between the growth rates in the [111],
[100], [110] directions becomes greater. Thus at 200°C in the range
$\Delta S = 0.2-0.8$ wt.% the ratio $\bar{v}_{[111],[100]}/v_{[110]}$ equals ~4, at 250°C
it equals 5 and at 300°C almost 6. Thus if the average growth rate
in the [111] and [100] directions is $\bar{v}_{[111],[100]} = k_1\Delta S$, and the growth
rate in the [110] direction is $v_{[110]} = k_2\Delta S$, at constant temperature
$k_1/k_2 = $ const.

On the basis of the data relating to the growth of sodalite
crystals and the solubility of sodalite under hydrothermal conditions, we may distinguish a metastable region on the concentration-
temperature diagram for $C_{NaOH} = $ const $= 30$ wt.%, within which
the sodalite crystals grow easily. In Fig. 15 the metastable region
(II) is delimited by two curves. The lower curve is that of the
saturation of the aqueous solutions (solubility curve), the upper
curve (broken line) separates the region in which parasitic crystals
start growing. The narrow zone within the metastable region above
the solubility curve indicates the region of critical supersaturations,
below which no growth of sodalite crystals takes place.

Surface of Dissolution of
Sodalite Crystals

After quenching, the surfaces of the seed blocks of sodalite were studied in a MIM-8M metallographic microscope. The surfaces of dissolution of the faces of the rhombododecahedron (110), tetrahedron (111), cube (100), and trigontritetrahedron (211) were examined at T = 200-250°C for an NaOH concentration of 10-30 wt.%.

The faces of the rhombododecahedron are characterized by the most stable dissolution picture. The surfaces are covered by flat rhomboid relict hillocks with blunted corners (Fig. 16). Similar "rounded" rhombs of greater or lesser clarity are encountered on the dissolution surface of the (110) faces almost independently of the temperature of the experiment and the concentration of the solvent.

The form of the dissolution surface of the (111) faces is shown in Fig. 17. The surface is represented by diffuse layer-like pyramids (Fig. 17a); in a number of cases etch pits in the form of negative trigonal pyramids may clearly be seen (Fig. 17b).

The most typical dissolution surface of the (100) faces at T = 250°C is represented by relict hillocks of almost square shape (Fig. 18). There are frequently etch pits of the same shape on these square projections (Fig. 18c).

Fig. 16. Dissolution surfaces of the faces of the rhombododecahedron (× 160).

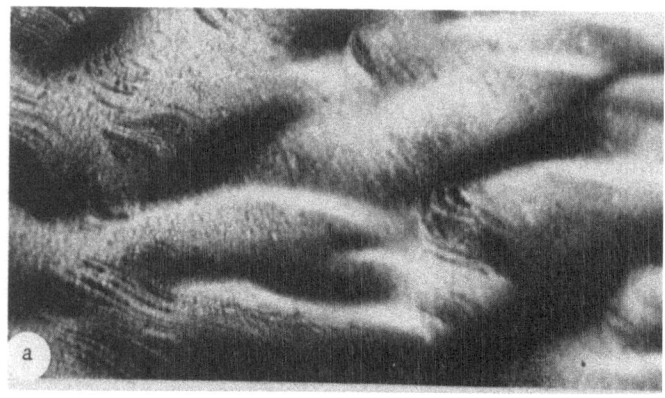

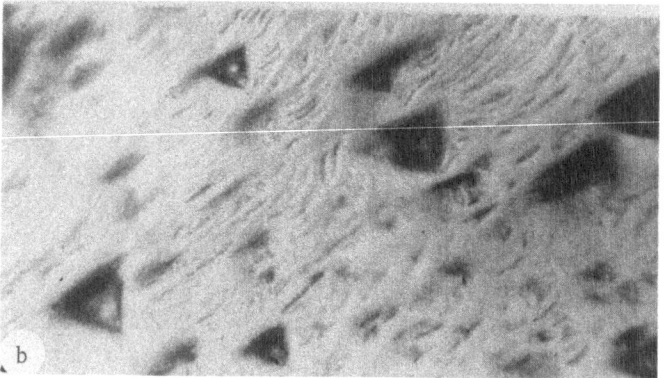

Fig. 17a, b. Dissolution surfaces of the faces of the tetrahedron
(× 170).

At 200°C the relief of the (100) faces is "softer" and is characterized by many fine etch pits surrounding square flat regions
(Fig. 18d). In individual regions of the (100) face disc-like, lamellar
dissolution figures appear. On the slopes of the "discs" occasional
deeper etch pits of extended form may clearly be seen (Fig. 18e).

In the dissolution of the (211) faces the surface is covered
with sharp hexahedral negative pyramids (Fig. 19a); the large
pyramids are intersected by shallow etch pits, also of hexahedral
shape (Fig. 19b), apparently associated with the outcropping
of dislocations on the surface of the face. It should be noted that
in this case the sodalite crystals were dissolved in a 10% solution
of NaOH, for which the stable solid phase was cancrinite, and the

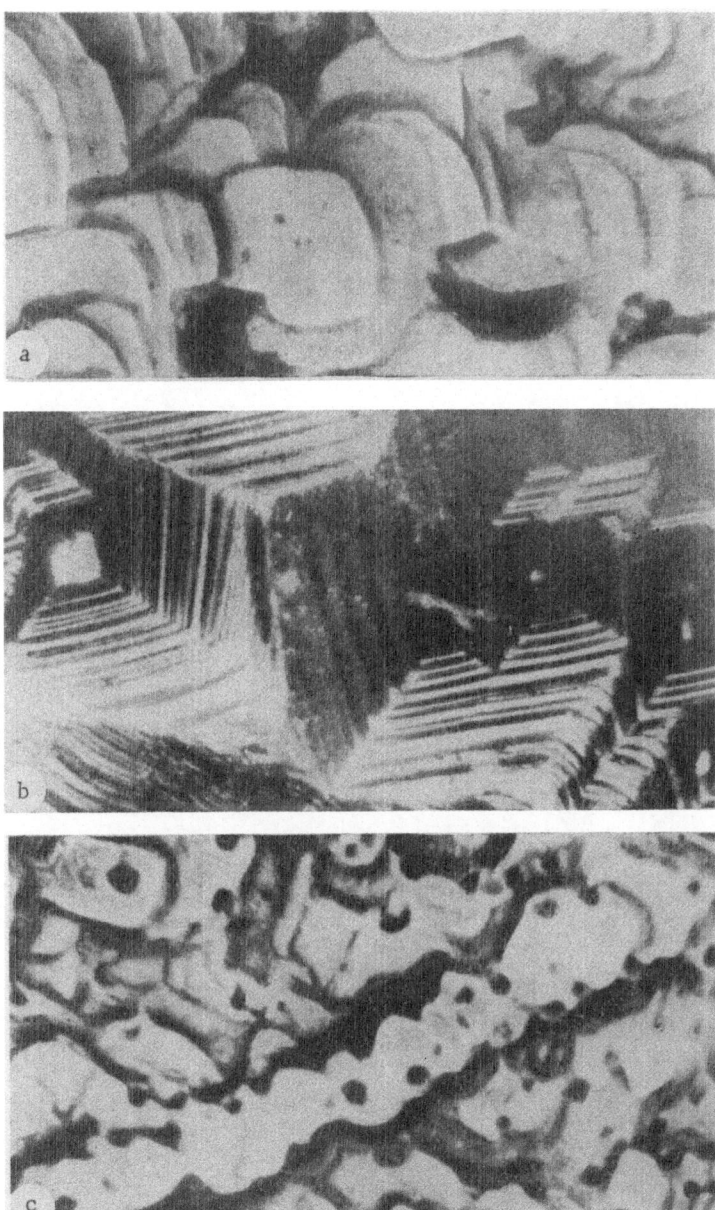

Fig. 18. Dissolution surfaces of the faces of the cube. a), b) × 160;
c) × 200.

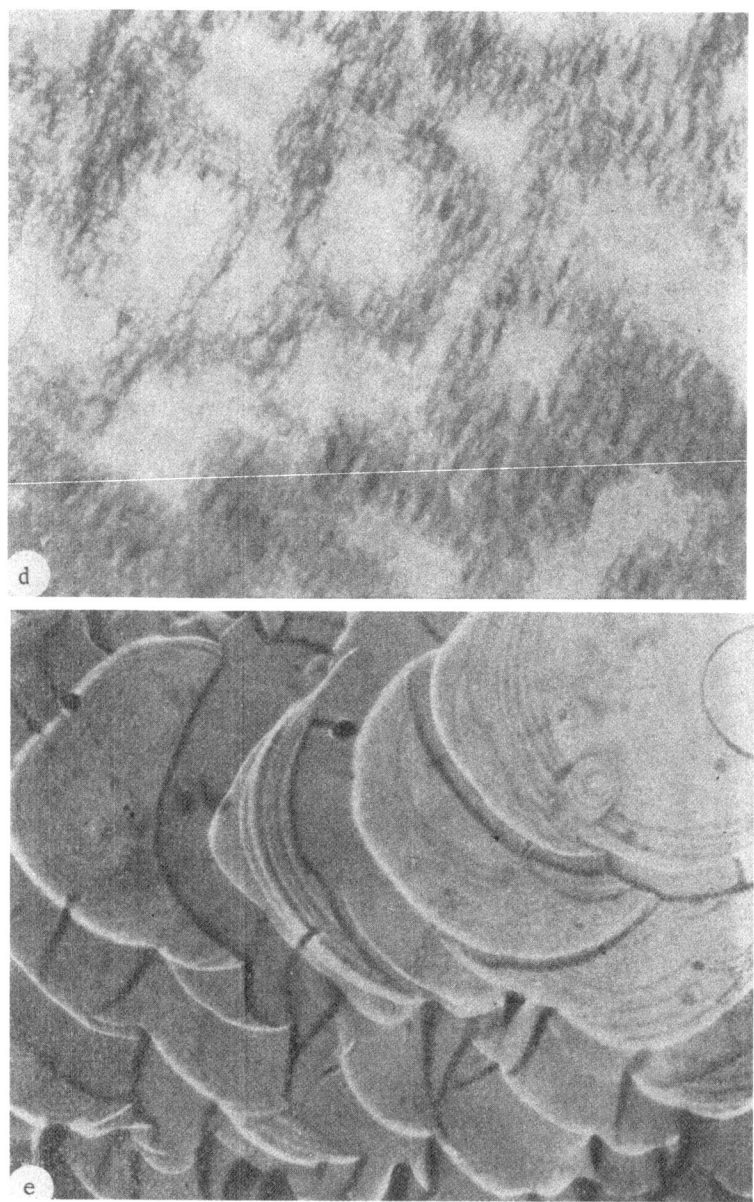

Fig. 18. (continued). d), e) × 170.

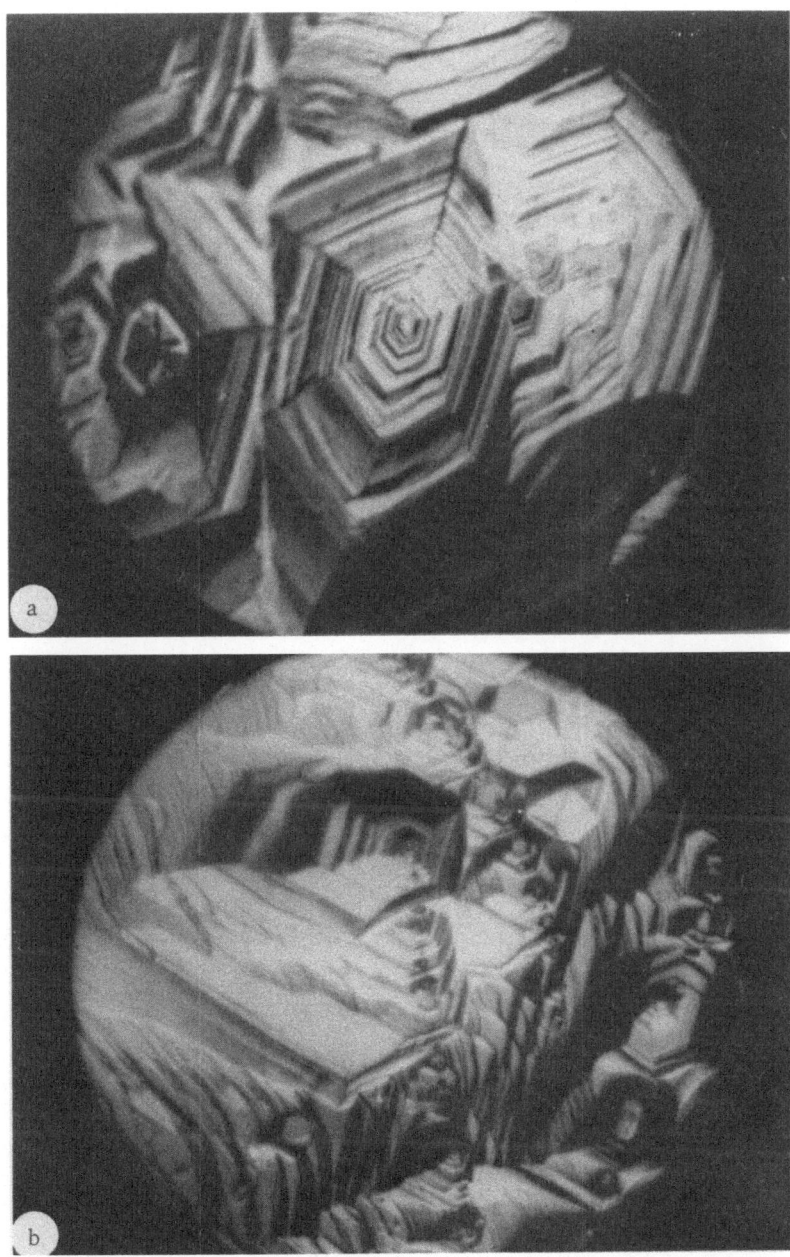

Fig. 19. Dissolution surfaces of the faces of the trigontritetrahedron (× 170).

sodalite crystals were covered with a deposit, an oriented over-growth of cancrinite, which was easily separated from the sodalite surface.

We see from the data presented that in the majority of cases the symmetry of the etch figures on the (100), (111), (110), and (211) faces reflects the symmetry of the face.

Conclusions

1. We have studied the solubility of natural and artificial sodalite in aqueous solutions of NaOH (up to 40 wt.% in concentration) between 200 and 300°C for an occupation factor of 0.8-0.9.

2. In the P, T, and C range studied, the solubility of sodalite is characterized by a positive coefficient. Increasing the NaOH concentration at constant temperature leads to a rise in the solubility of the sodalite.

3. We have calculated the heats of dissolution of sodalite in the temperature range 200-300°C for NaOH concentrations of 10, 20, and 30 wt.%.

4. The supersaturation in the growth zone varies linearly with ΔT in the temperature range studied.

5. We have considered the dissolution surfaces of the (100), (111), (110), and (211) faces.

6. The solubility data here obtained are related to the growth kinetics of sodalite single crystals under hydrothermal conditions. We have found a relation between the growth rates in the directions [111], [100], [110] and the supersaturation at constant temperature.

7. At constant temperature the ratio of the average growth rates of the rapidly-growing (111) and (100) faces to the growth rates of the rhombododecahedral (110) face remains constant over the range of supersaturations ΔS studied. On raising the temperature the ratio $\bar{v}_{[111]}$, $v_{[100]}/v_{[110]}$ becomes greater.

8. On the basis of data relating to the growth of sodalite crystals and their solubility under hydrothermal conditions, we have defined a metastable region, in coordinates of concentration versus temperature, for a constant NaOH concentration (30 wt.%), in which sodalite single crystals may be expected to grow.

Literature Cited

1. Yu. V. Shaldin, O. K. Mel'nikov, V. V. Nabatov, and Yu. V. Pisarevskii, Kris-
    tallografiya, 10:574 (1965).
2. Z. I. Zonn and I. S. Yanchevskaya, Zh. Neorg. Khim., 7:9 (1962).
3. L. V. Bryatov, "Some problems relating to the solubility and growth of quartz
    crystals," Candidate's Dissertation (in Russian), IK AN SSSR, Moscow (1955).
4. O. F. Tuttle and I. I. Friedman, J. Amer. Chem. Soc., 70:919 (1948).
5. A. F. Frederickson and I. E. Cox, Amer. Mineral. 39:738 (1954).
6. G. C. Kennedy, Econ. Geol., 39:25 (1944).
7. B. S. Krumgal'z and V. I. Mashovets, Zh. Prikl. Khim., 37:2596 (1964).
8. B. S. Krumgal'z and V. P. Mashovets, Zh. Prikl. Khim., 37:2750 (1964).
9. M. P. Vukalovich, Thermodynamic Properties of Water and Water Vapor [in
    Russian], Mashgiz (1955).
10. A. F. Frederickson and I. E. Cox, Amer. Mineral., 39:886 (1954).
11. R. A. Laudise and A. A. Ballman, J. Phys. Chem., 65:1396 (1961).
12. R. L. Barns, R. A. Laudise, and R. M. Shields, J. Phys. Chem., 67:835 (1963).

Crystallization of Sodalite on a Seed

O. K. Mel'nikov, B. N. Litvin, and N. S. Triodina

Crystallization under hydrothermal conditions takes place over a wider range of temperatures (100-800°C) and pressures (1-3500 atm) than crystallization from ordinary aqueous solutions (which occurs at temperatures no greater than 120-150°C), and the corresponding kinetic investigations may according be readily carried out. Another characteristic of crystallization under hydrothermal conditions is the constancy of the manner in which the supply material is fed into the growth zone. This takes place practically independently of the temperature and pressure and enables us to develop a theory of TC diagrams for hydrothermal systems [1].

1. Experimental Method

In order to study the growth of the faces of sodalite we used unlined autoclaves 1000 cm^3 in volume (H/D = 10) (Fig. 1). The temperature was measured with a chromel—alumel thermocouple placed in a special thermometric "pocket." The thermocouple was able to move freely in the "pocket," so that the temperature distribution with respect to the height of the autoclave could readily be measured. A diaphragm with a free area of $\Delta S \approx 5\%$ was placed halfway up the autoclave. As solvent we used a aqueous solution of NaOH with a constant concentration of 30% (d = 1.330). The charge was finely-divided hackmanite (a sulfur-containing form of sodalite), the amount of this being the same (200 g) for each experiment.

The samples used for the seeds were obtained by the growth of artificial material on large pieces of natural hackmanite at

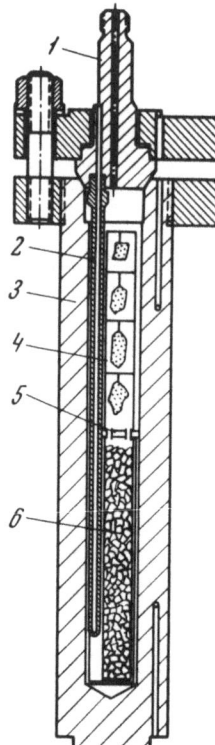

Fig. 1. Autoclave with thermometric
"pocket" for growing sodalite crystals.
1) Obturator; 2) thermometric pocket
with thermocouple inserted; 3) body of
autoclave; 4) frame with seeds; 5)
diaphragm; 6) charge.

300-350°C with ΔT = 20-30°C and a pressure of 400 atm. Rhombododecahedral faces formed on the samples in a few days. As
seeds we used plates 3-5 mm thick cut parallel to the (110), (100),
and (111) faces. Since the (100) and (111) faces usually failed to
appear on the faced samples, the plates were cut perpendicularly
to the three- and four-fold axes respectively. Four seeds were
placed in the reactor, the upper and lower ones being oriented in
the same way, thus enabling the temperature distribution in the
growth zone to be monitored. The point of this was that, in the
absence of a slope on the T–H curve (Fig. 2), the growth rates of
the upper and lower seeds should, in general, be the same. The
optimum temperature distribution was created by insulating the
upper part of the furnace and eliminating loss of heat.

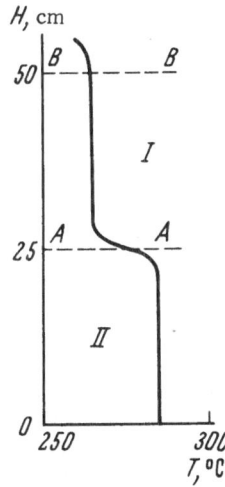

Fig. 2. Temperature distribution inside the autoclave. I) Growth zone; II) dissolution zone; A–A diaphragm; B–B closure.

The temperature of the experiments was varied from 150 to 400°C in steps of 50°C.* The supersaturation in the growth zone was varied by varying the temperature drop, equal to $\Delta T = 10$, 20 and 30°C. At 150°C a single experiment was carried out; it showed that in this case the growth rates were extremely low. No experiments were carried out above 400°C owing to the substantial mass transfer taking place.

The autoclave was heated to the specified working point without creating any temperature drop and as quickly as possible. On reaching the specified temperature the desired temperature drop was created in the autoclave. The duration of the experiments averaged 120 h–240 h at 150°C and 24 h at 400°C.

After the completion of an experiment the autoclave was disconnected and some of the solution was taken off through a valve so as to prevent the etching of the surface of the growing crystals.

The growth rate of the crystals was measured under MBS–1 binoculars using a micrometer eyepiece. The results of the measurements are all presented in Table 1.

---

*Subsequently we shall everywhere give the temperature of the growth zone.

TABLE 1. Growth Rates of

Face	$T = 160°$, $\Delta T = 20°$	$T = 200°$, $\Delta T = 10°$	$T = 200°$, $\Delta T = 15°$	$T = 200°$, $\Delta T = 20°$	$T = 200°$, $\Delta T = 30°$	$T = 250°$, $\Delta T = 10°$	$T = 250°$, $\Delta T = 20°$
(110)	0.010	0.006	0.017	0.022	0.047	0.060	0.14
	0.011	0.010	0.025	0.040	0.050	0.040	0.17
	0.012	0.014	0.016	0.040	0.050	0.058	0.19
	0.013	0.012	0.020	0.028	0.048	0.050	0.15
	0.014	0.016	0.019	0.020	0.045	0.050	0.16
	0.012	0.010	0.022	0.020	0.055	0.042	0.15
v_{av}	0.012	0.011	0.020	0.028	0.048	0.050	0.16
$k \cdot 10^2$	0.06	0.11	0.13	0.14	0.16	0.5	0.8
(100)	0.028	0.033	0.066	0.100	0.170	0.22	0.70
	0.020	0.033	0.072	0.100	0.170	0.23	0.66
	0.024	0.028	0.066	0.110	0.165	0.23	0.68
	0.028	0.040	0.078	0.090	0.185	0.22	0.65
	0.025	0.022	0.066	0.100	0.172	0.22	0.62
	0.025	0.028	0.066	0.100	0.170	0.25	0.64
v_{av}	0.025	0.031	0.067	0.100	0.172	0.22	0.66
$k \cdot 10^2$	0.12	0.31	0.4	0.5	0.54	2.2	3.3
(1$\bar{1}$1)	0.031	0.044	0.083	0.106	0.194	0.22	0.67
	0.037	0.055	0.083	0.110	0.210	0.20	0.61
	0.031	0.040	0.083	0.120	0.188	0.18	0.62
	0.031	0.050	0.083	0.115	0.192	0.16	0.64
	0.025	0.033	0.083	0.110	0.190	0.14	0.60
	0.031	0.056	0.083	0.108	0.186	0.20	0.60
$v_{(111)}$	0.031	0.045	0.083	0.116	0.190	0.18	0.62
(111)	0.025	0.040	0.075	0.130	0.210	0.25	0.65
	0.019	0.046	0.070	0.120	0.200	0.20	0.60
	0.025	0.050	0.067	0.125	0.210	0.18	0.58
	0.031	0.048	0.070	0.130	0.200	0.24	0.61
	0.025	0.052	0.073	0.120	0.190	0.23	0.59
	0.025	0.054	0.083	0.120	0.190	0.25	0.60
$v_{(111)}$	0.025	0.048	0.073	0.124	0.200	0.22	0.60
v_{av}	0.028	0.046	0.078	0.120	0.195	0.20	0.61
$k \cdot 10^2$	0.14	0.46	0.5	0.6	0.6	2	3

Sodalite Faces (in mm)

$T = 250°$, $\Delta T = 30°$	$T = 300°$, $\Delta T = 10°$	$T = 300°$, $\Delta T = 20°$	$T = 300°$, $\Delta T = 30°$	$T = 350°$, $\Delta T = 6°$	$T = 350°$, $\Delta T = 10°$	$T = 350°$, $\Delta T = 20°$	$T = 400°$, $\Delta T = 3°$
0.28	0.11	0.33	0.56	0.20	0.25	0.60	0.75
0.25	0.16	0.30	0.50	0.18	0.32	0.52	0.80
0.22	0.15	0.23	0.48	0.25	0.21	0.58	0.90
0.20	0.17	0.35	0.42	0.22	0.23	0.66	0.72
0.30	0.17	0.24	0.44	0.24	0.30	0.71	0.88
0.26	0.16	0.33	0.48	0.20	0.31	0.54	0.78
0.25	0.15	0.30	0.48	0.21	0.23	0.60	0.80
0.8	1.5	1.5	1.6	3.0	2.8	3.0	26.6
1.10	0.83	1.83	2.80	1.45	2.28	4.74	2.80
1.15	0.85	1.80	2.90	1.40	2.20	4.75	2.90
1.12	0.80	1.78	3.00	1.56	2.38	4.60	2.86
1.08	0.82	1.85	2.70	1.54	2.40	4.80	2.74
1.10	0.83	1.83	2.80	1.48	2.30	4.85	2.88
1.07	0.84	1.82	2.70	1.46	2.36	4.68	2.84
1.10	0.83	1.82	2.80	1.43	2.32	4.76	2.84
3.3	8.3	9.5	9.3	2.4	23.2	23.8	95
1.00	0.66	1.60	2.75	1.16	1.94	4.38	1.82
1.03	0.70	1.70	2.50	1.26	1.98	4.32	1.90
1.01	0.76	1.74	2.55	1.38	2.16	4.57	1.74
0.98	0.74	1.72	2.45	1.20	2.10	4.63	1.94
0.97	0.76	1.68	2.75	1.28	2.05	4.40	1.70
1.00	0.72	1.70	2.50	1.30	2.08	4.54	1.80
1.00	0.72	1.67	2.60	1.26	2.02	4.48	1.81
0.93	0.70	1.65	2.60	1.27	1.98	4.46	1.90
0.96	0.66	1.63	2.70	1.26	2.17	4.50	1.98
0.99	0.66	1.70	2.65	1.30	2.16	4.58	1.82
1.00	0.68	1.70	2.45	1.34	2.10	4.60	1.80
1.03	0.72	1.75	2.60	1.27	2.03	4.64	1.96
1.01	0.66	1.72	2.50	1.38	2.12	4.62	1.90
0.99	0.68	1.7	2.60	1.30	2.08	4.57	1.89
0.99	0.70	1.63	2.60	1.23	2.05	4.52	1.85
3.3	7	8.4	8.6	22	21	22.6	61

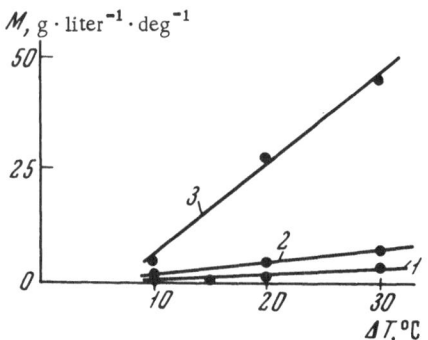

Fig. 3. Dependence of mass transfer on temperature drop: 1) 200°; 2) 250°; 3) 300°C.

2. Study of Mass Transfer

Owing to the fact that the reaction space of the autoclave
was divided into dissolution and crystallization zones, and also
owing to the existence of convection, a continuous mass-transfer
process occurred. The amount of material transferred to the
growth zone in unit time (referred to unit volume of solvent) may
be expressed in the form of a graphical relationship relative to
the growth parameters. The mass transfer M in the autoclave will
be determined by the rate of convection of the solution, which in
turn is a function of the temperature drop ΔT and the viscosity of
the solution. The mass transfer will also be determined by the
area and direction of the growing faces of the crystal.

We see from Fig. 3 that the mass transfer taking place
during the crystallization of sodalite varies linearly* with ΔT.
For a certain critical supersaturation ΔT_{cr} the mass transfer
equals zero. Taking values of 5, 4, and 2°C for ΔT_{cr} at 200, 250,
and 300°C and a value of almost zero at 350°C, we arrive at an
equation of the type:

$$M = \alpha_m (\Delta T - \Delta T_{cr}),$$

where α_m is a proportionality factor with dimensions of $g \cdot liter^{-1} \cdot$

---

*ΔT depends linearly on the supersaturation at constant temperature [2].

$deg^{-1} \cdot day^{-1}$. From the change taking place in the slope of the straight lines $M = f(\Delta T)$ at various temperatures we may easily establish the temperature dependence of the proportionality factor α_m:

$$\alpha_m = A \exp(-C/RT),$$

which coincides with the ordinary Arrhenius temperature dependence of the velocity constant of chemical reactions. Thus the mass transfer—crystallization temperature relationship may be reasonably accurately described by

$$M = A \exp(-C/RT)(\Delta T - \Delta T_{cr}).$$

The mass transfer also depends on the area of the seeds and apontaneously-formed crystals. We may therefore calculate the specific mass transfer $M_0 = M/S$, where S is the area of the seed surface. When spontaneous crystals are formed, it is difficult to estimate the growing area, and we are really only justified in speaking of specific mass transfer in connection with crystallization on seeds.

The duration of the experiments was chosen so as to leave part of the charge in the dissolution zone after the experiment. This enabled us to calculate the rate of charge consumption F per unit time referred to unit volume of the solution $(g \cdot liter^{-1} \cdot day^{-1})$ (Table 2). The consumption of charge increases very slightly in the range 200-250°C (Fig. 4). AT 260-280°C the $F = f(T)_{\Delta T = const}$ curves suffer a sharp break and almost coincide. This may be because the solubility is the same for the same temperature of the dissolution zone, and the consumption of the charge is independent of the temperature of the growth zone. The consumption of charge should vary with temperature in the same way as the solubility, and in absolute magnitude the comsumption of charge should exceed the solubility by the amount of material transferred to the growth zone. However, in the range 200-300°C, the charge consumption calculated from the weights of the residues lies well below the solubility values at the corresponding temperatures [2]. This discrepancy may be explained by the fact that, for low growth rates in the temperature range 200-250°C, not all the dissolved sodalite is used in growing the crystals on the seeds, the excess

TABLE 2. Results of Experiments on the Mass Transfer of Sodalite under Hydrothermal Conditions

Experiment	T, °C	ΔT, °C	Period of experiment, days	Weight of seed before experiment, g	Weight of seed after experiment, g	Weight of crystals formed spontaneously, g	Weight of material transferred in one day (mass transfer M), g/liter	Weight of residual charge, g	Weight of material expended in one day (consumption F), g/liter
1	160	20	4	24.2	24.7	—	0.12	195	1.41
2	200	10	9	19.7	22.4	—	0.37	180	2.75
3	200	15	6	21.8	26	—	0.87	172	5.9
4	200	20	5	30.8	35.7	—	1.25	160	10.0
5	200	30	3.5	29.8	40.6	—	3.75	164	12.9
6	250	10	5	13.2	19	—	1.55	180	5.33
7	250	20	2	19.9	33.9	—	9.35	180	13.3
8	250	30	5	17.1	50.2	10	11.5	115	22.6
9	300	10	3	20.1	34.1	5	9.0	80	57.2
10	300	20	3	14.1	52.4	20	27.7	40	7.56
11	300	30	3	14.5	59	50	45.0	20	85.7
12	350	6	2.7	15.5	75.6	10	40.0	50	84.7
13	350	10	2	24.9	50.3	40	50.3	30	131.0
14	350	20	2	20.0	88.6	70	106.0	20	139.0
15	400	3	1	18.3	34.0	5	34.5	50	83.3

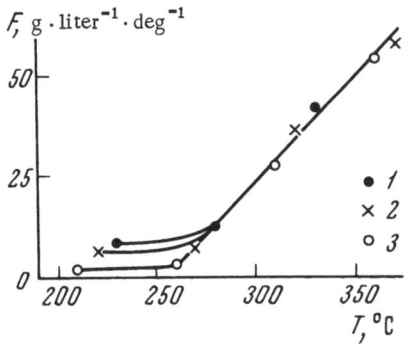

Fig. 4. Temperature dependence of the consumption of the charge. 1) $\Delta T = 30°$; 2) $20°$; 3) $10°$.

being crystallized within the charge on cooling (no growth takes place on the seeds during the cooling period, as after the autoclave has been disconnected the solution is taken out of the growth zone). This lack of agreement may also be partly explained by the long period required to establish equilibrium (4–5 days). At 300°C equilibrium is established in 10–12 h.

3. Kinetics of the Growth of Sodalite Faces

The measured growth rates (v) of the sodalite faces are given as functions of T and ΔT in Table 1. Figure 5 shows the dependence of v on ΔT at 200, 250, 300, and 350°C. Up to 200°C the order of the growth rates of the faces of sodalite is as follows: (111) > (100) > (110). Above 250°C the order becomes (100) > (111) > (110). The corresponding ratios of the growth rates of the (100), (111), and (110) faces are given in Table 3.

Over the whole range of temperature drops studied ($\Delta T = 10$–$30°C$) the relation between v and ΔT is linear, being described by the equation

$$v_{[hkl]} = k(\Delta T - \Delta T_{cr}) \quad \text{at} \quad \Delta T \geqslant \Delta T_{cr} ,$$

where k is a coefficient depending on the temperature and the crystallographic direction and having the dimensions of mm $\cdot$ deg^{-1}.

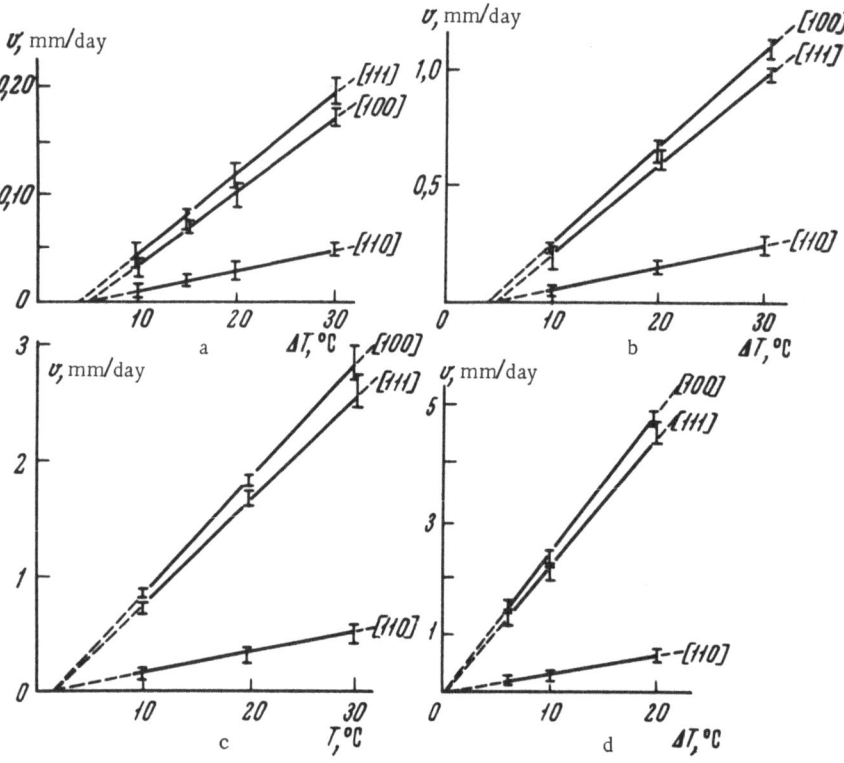

Fig. 5. Relation between the growth rates of sodalite crystals in the [100], [111], [110] directions and the temperature drop a) 200°; b) 250°; c) 300°; d) 350°C.

TABLE 3. Growth Rate Ratios of the Faces of Sodalite

T, °C	ΔT, °C	$v_{(100)}/v_{(1\cdot0)}$	$v_{(111)}/v_{(110)}$	T, °C	ΔT, °C	$v_{(100)}/v_{(110)}$	$v_{(111)}/v_{(110)}$
160	20	2.1	2.3	300	10	5.5	4.7
200	10	2.3	4.2	300	20	6.1	5.6
200	15	3.3	3.9	300	30	5.8	5.4
200	20	3.6	4.3	350	6	7.0	6.1
200	30	3.6	4.0	350	10	8.3	7.5
250	10	4.4	4.0	350	20	7.9	7.5
250	20	4.1	3.8	400	3	3.5	2.3
250	30	4.4	4.7				

day^{-1}, ΔT_{cr} is the critical temperature drop below which $v_{[hkl]} = 0$. The temperature dependence of ΔT_{cr} is indicated in Fig. 6. Above 350°C ΔT_{cr} tends to zero independently of the growth direction. Below 300°C the value of ΔT_{cr} for the (100) face increases more slowly than for the other two (111) and (110). In the range 200-250°C the curves for the (111) and (100) faces intersect owing to a change in the ratio of the growth rates of the faces of the cube and tetrahedron.

The temperature dependence of the growth rates of the (100) and (111) faces (Fig. 7) is excellently described by the equation

$$v_{[111], \ [100]} = A \exp (- B/RT) (\Delta T - \Delta T_{cr}).$$

For the slowly-growing (111) face we have a more complicated relationship:

$$v_{[110]} = A \exp \{- (B/RT + C/RT^2)\} (\Delta T - \Delta T_{cr}).$$

4. Morphological Characteristics of the Growth of the Faces

The face of the crystal provides the surface on which the whole crystallization process is performed. The small- and large-scale topography of this surface should closely reflect all the special characteristics of the process taking place on it. Yet we still have to discover how best to determine the laws relating

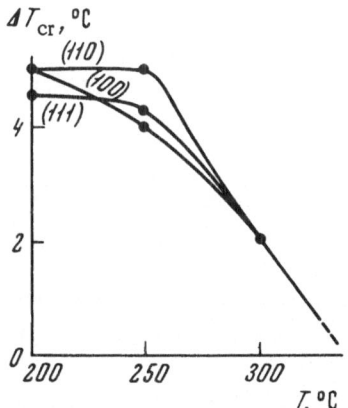

Fig. 6. Temperature dependence of ΔT_{cr}.

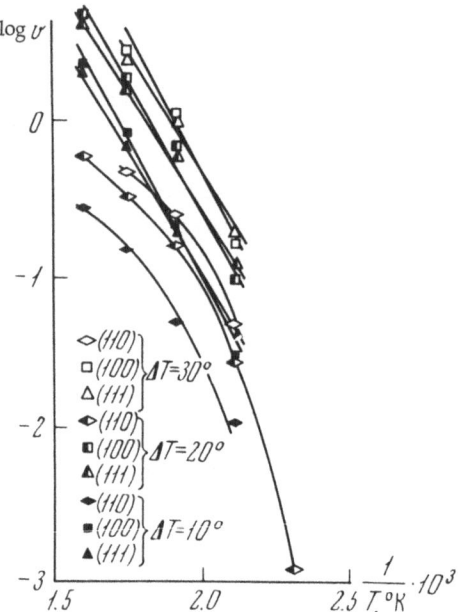

Fig. 7. Temperature dependence of the growth
rates of the (110), (100), and (111) faces.

the morphology to the growth mechanism and how this reflects
the quality of the growing layer. We carried out a number of
investigations with a view to solving this problem. We shall
now present some experimental material as to the macro- and
micromorphology of the growing faces of sodalite for strictly
specified conditions of growth.

After each experiment the sodalite single crystals were
washed with ethyl alcohol and studied visually in the MIM-8M
microscope. The sample quality was initially checked under a
binocular microscope.

On the (110) face we found several growth centers having the
form of stepped pyramids with a conical, or much more rarely a
plane top. Frequently the face accommodated no more than a
single growth center from which terraces several millimeters
wide propagated, these having a smooth surface. This picture appeared
on the faces situated in the lower part of the crystal. The faces
turned upward were covered with many small growth pyramids

banded by finer terraces, frequently overthrusting each other. Sometimes on the wide terraces (on the flat vertices or directly on the slope of the pyramid) new and finer growth pyramids appeared. Figure 8 shows a (110) face with one large growth center having a flat vertex on which many fine growth accessories have developed.

The (111) face (Fig. 9) is characterized by "many-headed" growth expressed in the form of a severely disjointed relief with acute, irrational faces. On the slopes of these faces no visible relief is to be found. The illustration clearly shows the face of the hexatetrahedron (321), rarely observed in sodalite crystals, with a unique undulating relief. The crests of the "waves" are quite parallel and extend over the whole face. The sides of the "waves" are mat, without any marked trace of relief.

The (100) face usually accommodates no particular growth centers (Fig. 10); it is covered with small flat cones or irregular hillocks. The sides of the hillocks often have a radially-disjointed relief. The uneven chain of cones or hillocks forms linear echelons which "proceed" from one of the edges of the face.

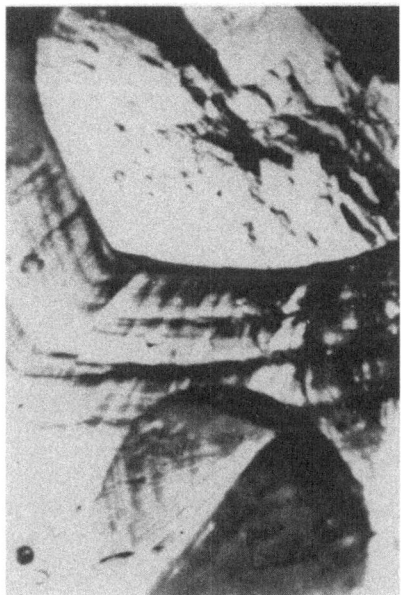

Fig. 8. Growth accessories on the (110) face for T = 300°C, $\Delta T = 30°C$ ($\times$ 15).

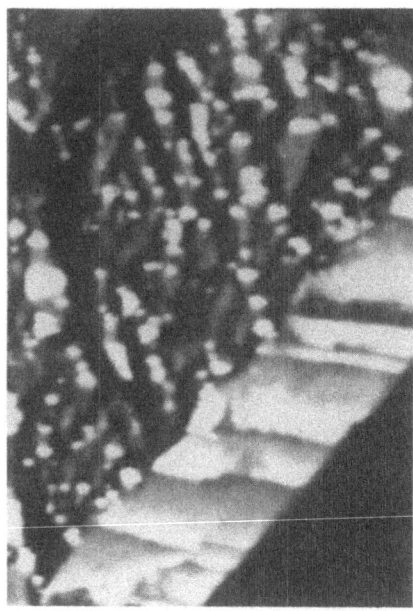

Fig. 9. Surface of the (111) face for T =
250°C, $\Delta T = 10°$ C ($\times$ 15).

Fig. 10. Surface of the (100) face for
T = 200°C, $\Delta T = 15°$ C ($\times$ 15).

The relief of the faces changes when changes take place in either the supersaturation or the temperature of the experiment. Figure 11 shows some photographs of the (110), (100), and (111) faces at 200°C for various ΔT values. In the case of low supersaturations the (110) face is characterized by gemoetrically regular rhomb-like pyramids with many stepped growth rings. With increasing supersaturation the character of the rings changes; they become more rounded, and secondary accessories appear on the sides (for $\Delta T = 20$°C).

Finally a plateau with a large number of growth centers in the form of rounded hillocks develops instead of the rhomboid cone (at $\Delta T = 30$°C).

For a small supersaturation no sharp relief appears on the (100) face. Further increasing ΔT leads to the development of a "soft," low relief, this gradually becoming "refined" ($\Delta T = 30$°C).

At $\Delta T = 5-15$°C the (111) face is covered with sharp, craggy accessories which gradually soften ($\Delta T = 20$°C) and become similar to those lying on the side of the cone. On further increasing the supersaturation the relief is "refined," while still preserving the features of many-headed growth ($\Delta T = 30$°C).

Since the (111) and (100) faces grew over rapidly with increasing temperature, the influence of temperature on the growth process could only be studied in detail for one face: the rhombododecahedron. It should be noted that at high temperatures the external form of the accessories on the (111) and (100) faces changed very little, and was reminiscent in character of the pictures observed on these faces at 200°C with $\Delta T = 20-30$°C. For the (110) face, however, the change in the configuration of the growth accessories with changing temperature was quite easy to follow.

Figure 12 shows the growth hillocks of the (110) face for the lowest (160°C) and highest (400°C) temperatures. The clear symmetry of the growth hillock at low temperatures and its absence at high agrees closely with earlier results [3]. At 250°C the configuration of the accessories has an intermediate character.

The (110) faces exhibit four types of accessories (Fig. 13): i) plane-rhombic, characteristic of low supersaturation and low temperatures; ii) conical-rhombic, characteristic of low and medium temperatures and medium supersaturations; (iii) conical-stepped,

Fig. 11. Changes in the surface of growing (110), (100), and (111)
A) (110), B) (100), C) (111); a, e, i: ΔT = 10°C, b, f, j:

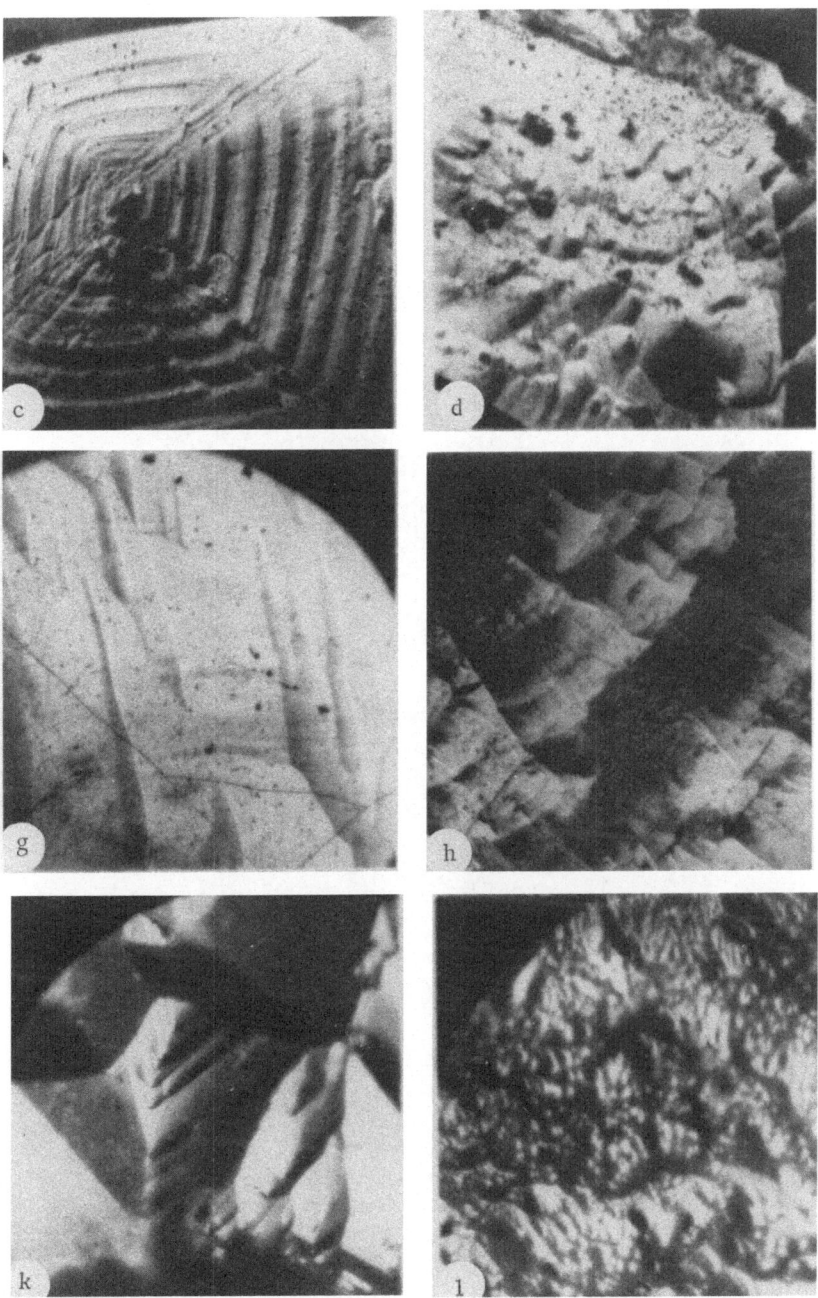

sodalite faces in relation to the temperature drop at 200°C (×100).
$\Delta T = 15°C$; c, g, k: $\Delta T = 20°C$, d, h, l: $\Delta T = 30°C$.

Fig. 12. Change in the growth accessories on the (110) face
at various temperatures (✗ 100). a) 160°C; b) 250°C; c)
400°C.

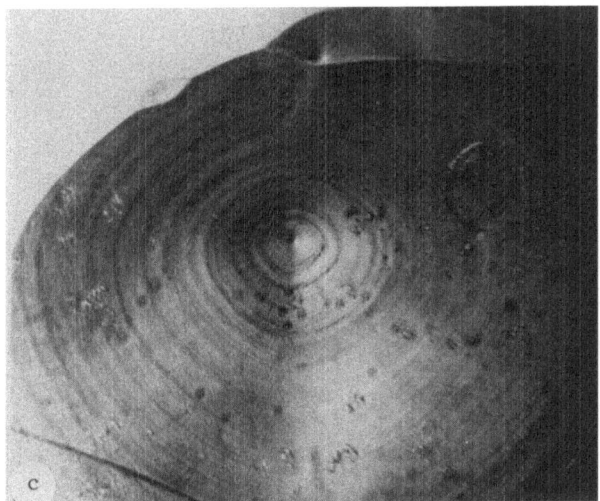

Fig. 12. (continued)

encountered together with the conical-rhombic accessories, but at higher temperatures and supersaturations; (iv) annular cones found in crystals grown at high temperatures and low supersaturations. Various intermediate cases occur between these four main types.

Many observations were made on the lateral surfaces of the accessories. The growth steps which are visible on almost all the accessories have end surfaces of different kinds. The plane-rhombic accessory has an end with a stepped relief; the profile may be represented in the form of concave steps (Fig. 14a). Sometimes (though rarely) the plane-rhombic accessories have a different type of end to the step, with a profile such as that shown in Fig. 14b. Such accessories develop for large supersaturations ($\Delta T = 30°C$). For slight supersaturations the end of the step on a plane-rhombic accessory has the form of a set of rectangular inclined surfaces.

Figures 14c and 14d show the ends of the steps on conical-stepped and conical-rhombic accessories, the latter being obtained under conditions of great supersaturation. The first of these exhibits a "seam" between the two ends of different steps.

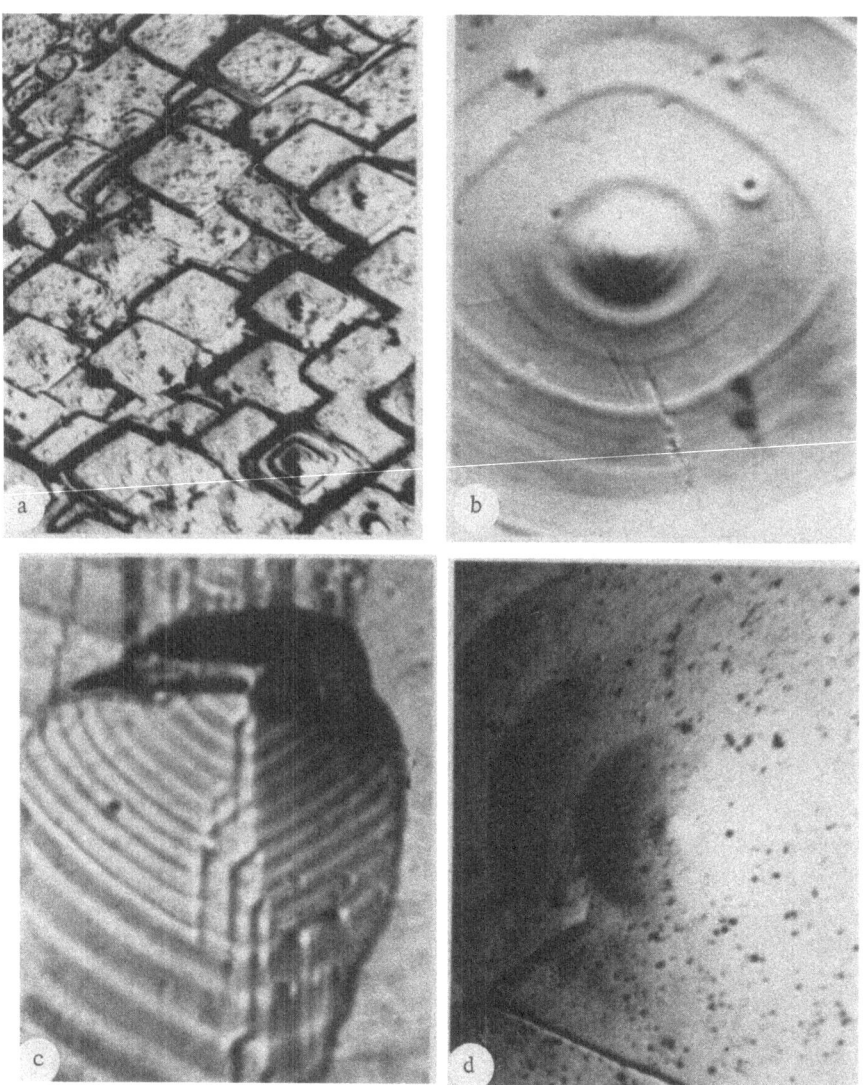

Fig. 13. Types of growth accessories encountered on the (110) face (× 200). a)
Plane-rhombic; b) conical-rhombic; c) conical-stepped. d) annular cones.

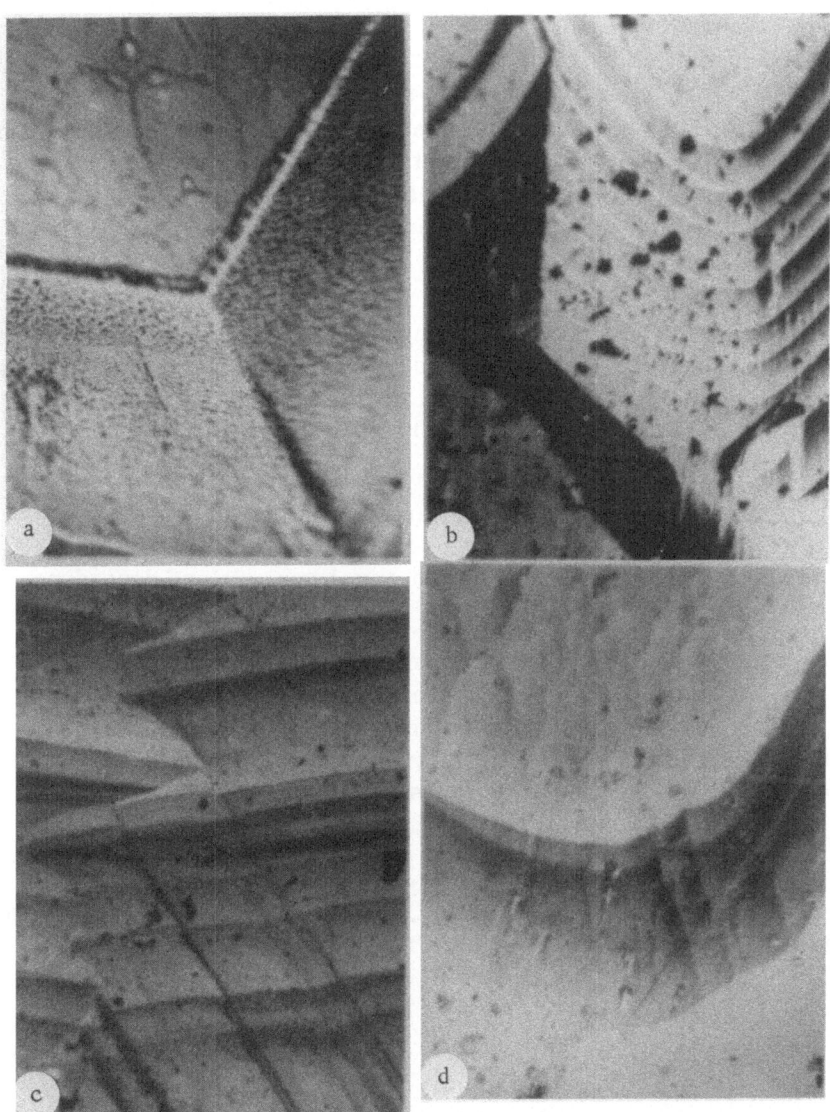

Fig. 14. Forms of steps in growth accessories on the (100) face (×150). a), b) on plane-rhombic accessories, T = 200°C, ΔT = 30°C; c) on conical-stepped accessories, T = 200°C, ΔT = 15°C; d) on conical-rhombic accessories, T = 300°C; ΔT = 10°C.

A study of the growth rates showed that the (110) face should have a growth mechanism differing from that of the (100) and (111); this is supported by visual examination of the macromorphology. Although we find no growth spirals directly on the surface of the (110) face, it is not impossible that the principal growth mechanism of the (110) face at small supersaturations will be of the dislocation type. An increase in supersaturation will probably lead to a quasi-three-dimensional, nucleation type of growth.

Conclusions

1. We have found that the mass transfer associated with the crystallization of sodalite depends linearly on the temperature drop for a constant growth temperature. The consumption of the charge increases very little in the temperature range 200-250°C, but it rises sharply at T = 260-280°C.

2. We have established a direct relationship between the growth rates of the sodalite faces and the temperature drop at constant temperature, and an exponential temperature dependence at constant supersaturation.

3. We have considered the macro- and micro-relief of the (110), (100), and (111) faces. With increasing supersaturation the growth figures on the (110) face change their polygonal forms into circular, while at large supersaturation "many-headed" growth occurs on the (100) and (111) faces.

Literature Cited

1.   B. N. Litvin and O. K. Mel'nikov, Kristallografiya, 13:5 (1968).
2.   L. N. Dem'yanets, E. N. Emel'yanova, and O. K. Mel'nikov, this volume, p. 125.
3.   F. C. Frank, Phil, Mag., 41:200 (1950).

Hydrothermal Crystallization of Lithium Silicates. Synthesis of Spodumene

V. A. Kuznetsov, A. A. Shternberg, and T. K. Ivanova

Crystallization in systems containing lithium, alumina, and silica has frequently been studied both from the point of view of producing individual crystals of lithium aluminosilicates (which include a number of compounds with low and even zero thermal-expansion coefficients) and also from that of studying the conditions underlying the genesis of natural lithium minerals. As a result of these investigations carried out in dry systems and also in the presence of water, almost all the natural lithium aluminosilicates have been synthesized, together with a number of artificial lithium compounds with no natural analogs. The only exception is α-spodumene, which has never been synthesized with an adequate degree of reproducibility. This would appear rather unexpected in view of the fact that spodumene, being a typical mineral of lithium-rich granite pegmatites, which are frequently formed in the final stages of the pegmatitic process, should not really require very high temperatures and pressures for its formation. Nevertheless, attempts at synthesizing α-spodumene up to 4000 atm and 600°C have usually proved unsuccessful [1-6]. Instead of α-spodumene, eucryptite, β-spodumene, petalite, and a number of artificial lithium compounds have been obtained. In individual experiments of Roy and Osborn [3] at 375-500°C and 840 atm, a phase interpreted as α-spodumene was obtained. Under similar conditions (360–400°C) Barrer and White obtained α-spodumene [2]. However, in both cases the products synthesized had lattice parameters differing from those of natural spodumene and higher refractive indices, indicating their contamination with impurities probably arising from the autoclave walls.

173

The important role of impurities in the crystallization of spodumene is supported by later investigations of Isaacs and Roy [5], who were unable to synthesize α-spodumene under "pure" conditions at pressures up to 4000 atm. At the same time, iron-containing spodumene was easily obtained by Scavnicar and Sabotier [6] at 550-600°C and ~1000 atm. An interesting observation was that of Shternberg [7] regarding the synthesis of chromium-containing spodumene at 2000-3000 atm and 700°C and the simultaneous (in the same ampoule) dissolution of natural colorless spodumene crystals.

The results of the foregoing investigations lead to the conclusions (a) spodumene is formed at pressures of over 4000 atm, as indicated, in particular, by the fact that all the phases synthesized have lower densities than that of α-spodumene, and (b) isomorphic impurities reduce the minimum pressure required for the formation of α-spodumene. The present investigation was aimed at verifying these points. We also made a more precise estimate of the conditions required for the formation of other lithium aluminosilicates, since comparison of the results of earlier investigations is frequently impeded by the fact that they were carried out with different alkalinities of the solutions, different pressures, and different materials in the original charge.

We took four basic variable parameters in our investigation: the alkalinity of the solution (pH), the temperature, the pressure, and the ratio of the components in the original mixture. The experiments were carried out in the temperature range 400-900°C at pressures of 750-9000 atm for solutions of pH 3-12. The lithium compounds were synthesized in solutions of LiCl or LiOH, using a combination of the oxides Al_2O_3 and SiO_2. The alkalinity of the solutions was varied by adding HCl to the original LiCl, or else LiOH. The pH of the solutions was measured immediately after opening the ampoules.

The experimental technique was as follows. The original charge (the sum of the SiO_2 and Al_2O_3 equaled ~1 g) was loaded into copper ampoules and the solutions was then poured in (2 cm³ of solution to 1 g of mixture). Between six and nine ampoules were placed in the autoclave at the same time, the maximum parameters being 3000 atm at 700°C [8]. Experiments at over 3000 atm were carried out in an special installation, the construction of which

TABLE 1. Synthesis of Lithium Aluminosilicates

Serial No.	Wt. ratio Al₂O₃/SiO₂ in the charge	Solutions	pH of solution	T_{exp}, °C	Pressure, atm	Results[*]
1	1 : 1	LiCl, 40%	~5	660—700	2000—3000	α-eucryptite in the form of short prismatic hexagonal crystals (~ 1 × 0.5 mm)
2	1 : 2	LiCl, 40%	~5	600—700	2000—3000	α-eucryptite in the form of short prismatic hexagonal crystals (2 × 1 mm), β-spodumene in the form of a dipyramid (~ 0.3 mm)
3	1 : 3	LiCl, 40%	~5	660—700	2000—3000	β-spodumene in the form of dipyramidal crystals
4	1 : 5	LiCl, 40%	~5	660—700	2000—3000	β-spodumene, quartz
5	1 : 8	LiCl, 40%	~5	660—700	2000—3000	Petalite in the form of lamellar crystals
6	1 : 2	LiCl, 60%	~5	600—700	2000—3000	α-eucryptite in the form of short prismatic crystals (1 × 0.5 mm), β-spodumene
7	1 : 2	LiCl, 20%	~5	600—700	2000—3000	α-eucryptite, β-spodumene
8	1 : 2	LiCl, 10%	~5	660—700	2000—3000	α-eucryptite, β-spodumene
9	1 : 2	LiCl + HCl	~3	660—700	2000—3000	α-eucryptite, β-spodumene
10	1 : 1	LiCl + LiOH, 40%	7—9	660—700	2000—3000	α-eucryptite in the form of acicular crystals (0.5 × 0.05 mm), lithium silicate in the form of needles
11	1 : 1	LiOH	> 11	660—700	2000—3000	α-eucryptite in the form of acicular crystals (0.5 × 0.02 mm), lithium silicate
12	1 : 2	LiCl + LiOH	7—9	660—700	2000—3000	α-eucryptite (acicular crystals), lithium silicate, β-spodumene
13	1 : 2	LiOH	> 11	600—700	2000—3000	α-eucryptite, petalite, lithium silicate (individual crystals)
14	1 : 3	LiCl + LiOH	7—9	660—700	2000—3000	α-eucryptite, lithium silicate, β-spodumene
15	1 : 3	LiOH	> 11	660—700	2000—3000	α-eucryptite, lithium silicate in the form of needles, petalite
16	1 : 5	LiOH	> 11	600—700	2200—3000	Lithium silicate, petalite in the form of needles, α-eucryptite
17	1 : 8	LiOH	> 11	660—700	2200—3000	fibrolitic crystals of petalite, lithium silicate
18	1 : 1	LiCl, 40%	~5	500—520	1800—2600	α-eucryptite
19	1 : 2	LiCl, 40%	~5	500—520	1800—2600	α-eucryptite, β-spodumene

TABLE 1. (Continued)

Serial No.	Wt. ratio Al_2O_3/SiO_2 in the charge	Solutions	pH of solution	T_{exp}, °C	Pressure, atm	Results*
20	1 : 3	LiCl, 40%	~5	500—520	1800—2600	β-spodumene, α-eucryptite
21	1 : 5	LiCl, 40%	~5	500—520	1800—2600	β-spodumene, quartz, individual crystals of petalite
22	1 : 8	LiCl, 40%	~5	500—520	1800—2600	Petalite in the form of flat crystals
23	1 : 2	LiCl + HCl	~3	500—520	1800—2600	β-spodumene
24	1 : 5	LiCl + HCl	~3	500—520	1800—2600	β-spodumene, quartz
25	1 : 8	LiCl + HCl	~3	500—520	1800—2600	β-spodumene, petalite
26	1 : 1	LiCl + LiOH	7—9	500—520	1800—2600	a-eucryptite in the form of acicular crystals, lithium silicate
27	1 : 1	LiOH	> 11	500—520	1800—2600	a-eucryptite in the form of acicular crystals, lithium silicate
28	1 : 2	LiCl + LiOH	7—9	500—520	1800—2600	a-eucryptite, lithium silicate
29	1 : 2	LiOH	> 11	500—520	1800—2600	Lithium silicate, a-eucryptite, petalite in the form of needles
30	1 : 3	LiCl + LiOH	7—9	500—520	1800—2600	a-eucryptite, petalite, β-spodumene (individual crystals)
31	1 : 5	LiCl + LiOH	7—9	500—520	1800—2600	a-eucryptite, petalite
32	1 : 5	LiOH	> 11	500—520	1800—2600	Lithium silicate, petalite (acicular crystals), a eucryptite
33	1 : 8	LiOH	> 11	400—420	1800—2600	Lithium silicate, petalite (acicular crystals)
34	1 : 1	LiCl, 40%	~5	400—420	1500—2400	β-spodumene, α-eucryptite
35	1 : 2	LiCl, 40%	~5	400—420	1500—2400	a-eucryptite, β-spodumene
36	1 : 5	LiCl, 40%	~5	400—420	1500—2400	a-eucryptite, β-spodumene,
37	1 : 8	LiCl, 40%	~5	400—420	1500—2400	Petalite
38	1 : 2	LiCl + HCl	~3	420	1500—2400	β-spodumene
39	1 : 8	LiCl + HCl	~3	420	1500—2400	β-spodumene, petalite
40	1 : 1	LiOH	> 11	400—420	1500—2400	a-eucryptite in the form of acicular crystals, lithium silicate
41	1 : 3	LiOH	> 11	400—420	1500—2400	Lithium silicate, petalite in the form of acicular crystals a-eucryptite (individual crystals)
42	1 : 5	LiOH	> 11	400—420	1500—2400	Lithium silicate, petalite in the form of needles
43	1 : 8	LiOH	> 11	400—420	1500—2400	Petalite, lithium silicate

*Phases shown in thicker type predominate. In other cases the phases are formed in roughly equal amounts.

TABLE 2. Synthesis of Spodumene

Serial No.	Charge	T_{exp}, °C	Pressure atm	Results
1	Al_2O_3, SiO_2 impurity Cr_2O_3	630	3000	Growth of colored spodumene on a natural seed. Synthesis of α-spodumene in the charge not observed
2	SiO_2, Fe_2O_3	400—680	250—3000	Iron-containing spodumene
3	SiO_2, Cr_2O_3	400—630	500—1000	Mixture of original oxides
4	SiO_2, Cr_2O_3	400—680	750—3000	Chromium-containing spodumene, quartz
5	Al_2O_3, SiO_2	700	5600	α-eucryptite, β-spodumene
6	Al_2O_3, SiO_2	700	5300	β-spodumene, α-eucryptite
7	Al_2O_3, SiO_2	700	6500	α-spodumene
8	Al_2O_3, SiO_2	800	6700	α-spodumene,
9	Al_2O_3, SiO_2	800	5700	β-spodumene, α-eucryptite
10	Al_2O_3, SiO_2	800	8100	α-spodumene
11	Al_2O_3, SiO_2	900	5300	β-spodumene
12	Al_2O_3, SiO_2	900	8100	α-spodumene

will be published later. The experiments lasted for 48-82 h, crystallization being entirely completed in this period. Altogether 200 experiments or more were carried out. The crystallization products were examined under the microscope and by x-ray diffraction.

The work was carried out in two stages. In the first stage we studied the crystallization of lithium aluminosilicates at pressures of up to 3000 atm (varying the temperature, the alkalinity of the solution, and the ratio of the components in the original charge); in the second stage we studied the crystallization of α-spodumene at pressures of over 3000 atm. The alkalinity of the solution and the Al_2O_3/SiO_2 ratio in the charge were kept constant.

The principal results of the experiments are presented in Tables 1 and 2.

Crystallization of Lithium Aluminosilicates

It follows from Table 1 that in pure LiCl solutions at 400-700°C and pressures of 750-3000 atm we may synthesize α-eucryptite $LiAlSiO_4$, β-spodumene β-$LiAlSi_2O_6$, and petalite $LiAlSi_4O_{10}$.* The

---

*In individual cases a mica-like compound was occasionally synthesized at 420-450°C. The composition and conditions of synthesis of this phase were not uniquely established.

α-eucryptite is mainly formed for high Al_2O_3/SiO_2 (A/S) ratios in the original charge ($\sim 1:1$), β-spodumene for smaller ratios ($1:2$ to $1:3$), and petalite for a high concentration of silica (ratio A/S $\sim$ $1:8$). In the neighborhood of the ratio A/S $\sim 1:5$ we observe paragenesis of β-spodumene and quartz.

The concentration of the LiCl solutions, and equally the temperature and pressure within the limits indicated, have little effect on phase formation in the $LiCl-Al_2O_3-SiO_2-H_2O$ system. Thus in a series of experiments we varied the concentration of the LiCl solution from 10 to 60% (the total amount of solution and the sum of the alumina and silica in the charge remained constant). The A/S ratio was $1:2$, which corresponded to the simultaneous crystallization of α-eucryptite and β-spodumene in a 40% solution of lithium chloride. The experiments showed that increasing the concentration of the LiCl solution to 60% had hardly any affect on the ratio of the phases, and only on reducing the concentration of the solution below $\sim 20\%$ did the amount of β-spodumene diminish and the α-eucryptite become dominant in the crystallization products.

The pH of the solution, together with the A/S ratio, constitutes the second most important factor in the crystallization. Reducing the alkalinity of the solution (in LiCl + HCl solutions) has practically no effect on the phase formation, but on increasing the alkalinity (in LiCl + LiOH or LiOH solutions) new compounds appear and the ranges of crystallization of the phases already noted alter. First of all lithium silicate Li_2SiO_3 appears. For a ratio of A/S = $1:1$ lithium silicate is synthesized for high values of the alkalinity, but with increasing concentration of SiO_2 in the original mixture the range of its stability moves in the low pH direction (Fig. 1). In alkaline solutions the field of formation of petalite also widens, and the latter is synthesized for higher A/S ratios ($\sim 1:2$ for pH > 10) than in LiCl solutions. The range of existence of β-spodumene, on the other hand, contracts slightly on increasing the alkalinity (Fig. 1), and for pH > 10 it becomes unstable, being replaced by α-eucryptite and lithium silicate.

Description of the Phases

The α- eucryptite $LiAlSiO_4$ formed at low and moderate pH values of the solutions (3-8) is represented by large crystals

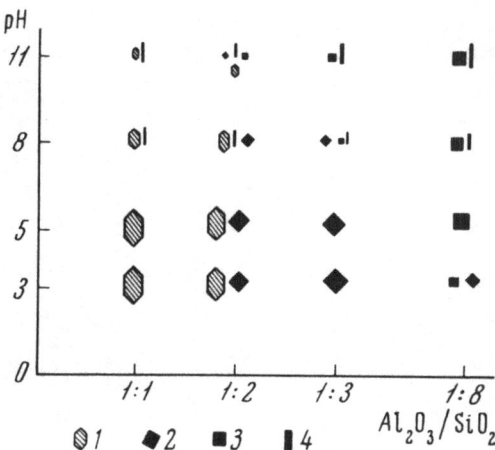

Fig. 1. Fields of stability of lithium silicates in relation to the pH of the solution and the ratio of the original charge components at T = 660-700°C and P = 2000-3000 atm. 1) α-LiAlSiO$_4$; 2) β-LiAlSi$_2$O$_6$; 3) LiAlSi$_4$O$_{10}$; 4) Li$_2$SiO$_3$.

(up to 2 mm) in the form of pseudohexagonal prisms, together with a bipyramid. Cruciform twins of α-eucryptite and druses of crystals growing together are often formed (Fig. 2). The crystals are usually colorless, transparent, and with an extremely perfect facing.

For solution pH values of over 8 the size of the α-eucryptite crystals diminishes to 0.1-0.6 mm and they assume the form of thin hexagonal prisms with an axial ratio of $c/a \sim 1:20$.

The diffraction pattern of our own α-eucryptite agrees with that of the latter given by Barrer and White [2].

The β-spodumene, β-LiAlSi$_2$O$_6$, is represented in every experiment by fine (0.2-0.3 mm) crystallites in the form of dipyramids (Fig. 3), and only very rarely does one encounter larger samples (up to 0.5 mm). Dense concretions of crystals are often formed and it is difficult to separate any individuals from these. The β-spodumene crystals are colorless and transparent, but contain many defects: cracks, inclusions of charge particles which have not completely reacted, etc.

Our x-ray diffraction pattern of the β-spodumene so synthesized agrees with that given in [4].

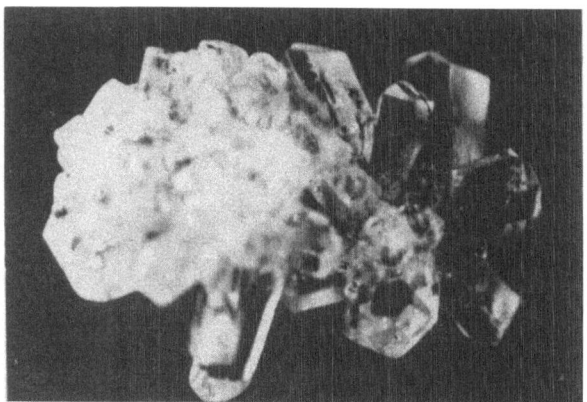

Fig. 2. Druse of α-eucryptite crystals (× 20).

Fig. 3. Crystals of β-spodumene (× 40).

Petalite, $LiAlSi_4O_{10}$, is represented by either platelets or
acicular crystals, depending on the alkalinity of the solution. For
pH = 5-8 (A/S ~ 1:8) tabular, colorless crystals up to 0.8 mm in
size are usually formed (in individual cases up to 0.5 mm) (Fig.
4a). Usually the crystals are severely twinned. In a strongly alkaline medium (pH > 10) the petalite crystals assume the form
of thin needles (up to 0.1 mm long) forming a cotton-wool-like mass
(Fig. 4b). The x-ray diffraction patterns of the crystals are iden-

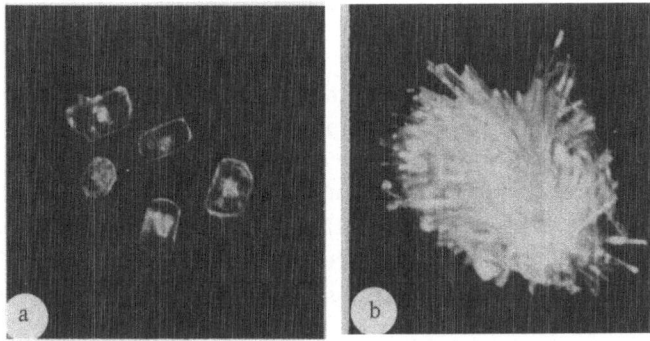

Fig. 4. Crystals of petalite (× 30). a) Tabular; b) acicular.

tical in both cases and correspond to the petalite x-ray pattern given in [3].

Lithium silicate, Li_2SiO_3, is only formed in an alkaline medium and always takes the form of long, brittle needles (Fig. 5), up to 1-3 mm long.

Of all the phases described lithium silicate crystallizes the most easily, and in individual cases the crystals reach greatest dimensions of up to 10 mm with a width of up to 5 mm.

The x-ray diffraction pattern of lithium silicate is identical with that of Li_2SiO_3 given by Barrer and White [2].

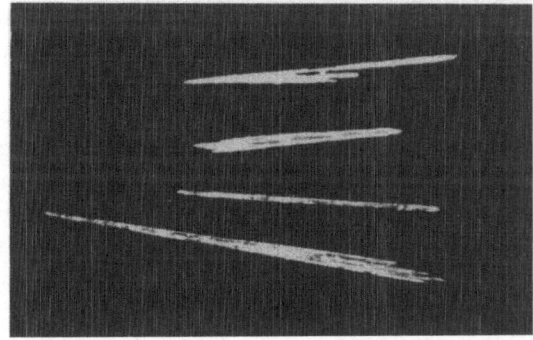

Fig. 5. Crystals of lithium silicate (× 20).

Synthesis of Spodumene

In none of the experiments just described was the formation of α-spodumene observed. Furthermore, natural spodumene crystals were placed in each ampoule, and by observing the qualitative changes in these we were able to judge the stability of spodumene in various conditions. We found that in weakly-acid and weakly-alkaline solutions, (pH = 5-8) the spodumene crystals hardly changed at all, but in more acid solutions they dissolved slowly, while in alkaline solutions considerable changes occurred; on the surface of the α-spodumene we observed crystals of α-eucryptite, β-spodumene, lithium silicate, and petalite, depending on the A/S ratio in the original mixture. These experiments clearly show that alkaline media are the least favorable for the synthesis of spodumene. In subsequent work we used pure solutions of lithium chloride. The A/S ratio in the charge was 1:2.

The results of our experiments agree with those of Isaacs and Roy [4] and support the conclusion that at pressures of under 4000 atm and in the absence of impurities no spodumene is of formed. The α-spodumene is the densest of all phases synthesized in our experiments (density of α-eucryptite $\sim$2.5 g/cm^3, of β-spodumene $\sim$2.4 g/cm^3, of petalite $\sim$2.45 g/cm^3, and of α-spodumene $\sim$3.15 g/cm^3), from which we may conclude that its stability range lies at much higher pressures. The formation of spodumene at lower pressures in the presence of impurities observed by a number of authors may be logically explained as being due to the isomorphic substitution of Al^{3+} by the larger cations, Fe^{+3}, Cr^{+3}, etc. ($r_{Al^{+3}} = 0.57$ Å, $r_{Fe^{+3}} = 0.67$ Å, $r_{Cr^{+3}} = 0.64$ Å).

In order to verify the foregoing assertion, in a number of experiments we partly or completely replaced the alumina in the original mixture by Fe$_2$O$_3$ or Cr$_2$O$_3$. We found that on partially replacing Al$_2$O$_3$ by Cr$_2$O$_3$ an overgrowth of colored spodumene took place on the natural cyrstal at 3000 atm and 700°C (Table 2, No. 1). The amount of Cr^{+3} in the surface layer was not determined quantitatively, however, the intense green coloring of the layer (thickness up to 1 mm) clearly indicated the isomorphic intrusion of Cr^3 into the crystal. Since no spodumene grew on the natural crystal under the same conditions but in the absence of Cr$_2$O$_3$, it is quite clear that the trace of Cr^{+3} "expanded" the range of stability of spodumene and enabled it to form at lower pressures.

The replacement of all the aluminum oxide by Fe_2O_3 or Cr_2O_3 in the mixture leads to the formation of iron on chromium spodumene at low pressures (Table 2). In the temperature range 400-700°C iron spodumene is easily synthesized at ~250 atm. Chromium spodumene is formed above ~750 atm. Both iron and chromium spodumenes comprise dark green acicular crystal up to 0.1 mm long.

In order to discover the conditions of formation of "pure" aluminum spodumene, α-$LiAlSi_2O_6$, we carried out some experiments at 700-800°C under pressures of up to 9000 atm. The A/S ratio in the original charge was 1:2.

In these experiments α-spodumene in the form of extended, compact crystals up to 1 mm long (Fig. 6) was synthesized at pressures of over 6000 atm (Table 2). At lower pressures α-eucryptite and β-spodumene were synthesized. The x-ray diffraction pattern on the synthetic α-spodumene agreed with the corresponding standard.

Conclusion

In the Li^+-Al_2O_3-SiO_2-H_2O system in the temperature range 400-700°C and at pressures of up to 9000 atm we have synthesized α-eucryptite, β-spodumene, petalite, lithium silicate, and α-spodumene over a wide range of pH values. The fields of stability of the

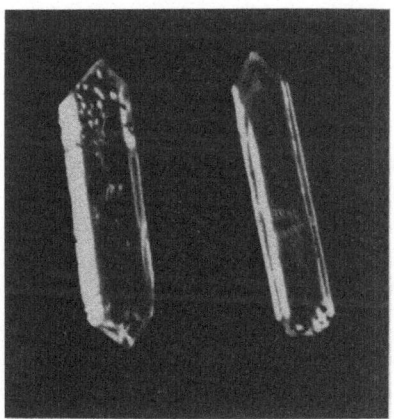

Fig. 6. Crystals of synthetic α-spodumene ($\times$ 4).

phases are mainly determined by the ratio of the components in the charge, the pH of the solution, and the pressure. Lithium silicate and petalite are more stable in an alkaline medium, but for high silica contents in the original mixture the latter is also formed at medium pH values (5-8). The range of existence of the β-spodumene, on the other hand, contracts with increasing alkalinity of the solution, while α-eucryptite is stable in both acid and alkaline media.

The α-spodumene is synthesized in LiCl solutions at pressures of over ~6000 atm. At lower pressures it decomposes into α-eucryptite and β-spodumene. The partial replacement of Al^{+3} by Cr^{+3} reduces the lower limit of pressures for the formation of spodumene; spodumene containing traces of Cr^{+3} grows on a seed at 3000 atm. The complete replacement of the Al_2O_3 by Fe_2O_3 or Cr_2O_3 leads to the synthesis of iron and chromium spodumene in the low-pressure range: ~250 and ~750 atm respectively.

Literature Cited

1.   R. A. Hatch, Amer. Mineral., 28:471 (1943).
2.   R. Barrer and E. White, J. Chem. Soc. London, 1267 (1951).
3.   R. Roy and E. F. Osborn, J. Amer. Chem.Soc., 71:2086 (1949).
4.   R. Roy, D. Roy, and E. F. Osborn, J. Amer. Cer. Soc., 33:152 (1950).
5.   T. Isaacs and R. Roy, Geochim. Cosmochim. Acta, 15:213 (1958).
6.   S. Scavnicar and G. Sabotier, Bull. Soc. Franc. Mineral. Crist., 80:308 (1957).
7.   A. A. Shternberg, in: Hydrothermal Synthesis of Crystals [in Russian], Izd. Nauka (1968), p. 203.
8.   V. A. Kuznetsov and A. A. Shternberg, Kristallografiya, 12:336 (1967).

Hydrothermal Synthesis of Germanates

I. P. Kuz'mina, B. N. Litvin, and V. S. Kurazhkovskaya

The limits of phase formation and crystallization in the $Na_2O-ZnO-GeO_2-H_2O$ system at high temperatures and pressures was discussed earlier [1]. The fundamental possibility of synthesizing germanates of various compositions under hydrothermal conditions was clearly indicated. In order to obtain a more comprehensive knowledge of the processes taking place in this complex system we initiated an investigation into phase formation in the $Na_2O-GeO_2-H_2O$, $Na_2O-ZnO-GeO_2-H_2O$ systems.

It has been suggested that potassium germanates cannot be obtained by crystallization from aqueous solutions [2]. This would appear to be true simply for fairly low temperature and pressures. We thought it interesting to study the synthesis of the $K_2O-GeO_2-H_2O$ and $K_2O-ZnO-GeO_2-H_2O$ systems under hydrothermal conditions as well.

The resultant experimental data enabled us to establish the boundaries limiting the regions of crystallization of germanates of different compositions. In the T and P range studied the temperature and pressure have no effect on the boundaries of the crystallization fields; the concentration of the solvent determines the ranges of existence of the various phases. The (T–C) diagram plotted in temperature–concentration coordinates is not the classical phase diagram, but it gives a fair representation of the phase relationships in the systems under consideration. A preliminary explanation of the T–C diagrams is given in [3].

Hydrothermal conditions of synthesizing various single crystals enable us to carry out chemical reactions at fairly high tem-

185

peratures and pressures, at which the solubility of the majority
of poorly-soluble substances in aqueous solutions of salts, acids,
or bases rises sharply. This in turn offers the possibility of
producing the desired substance in the form of a single-crystal
product.

In studying the conditions required for the formation of germanates of various compositions in the systems just mentioned
we used existing data relating to the dissolution and crystallization
of ZnO in aqueous solutions of NaOH and KOH [4].

As original components we used the chemical reagents
zinc oxide (chemically pure), caustic soda (chemically pure), caustic potash (chemically pure),and silicon dioxide (especially pure)
of the hexagonal type. The alkali solutions were stored in polyethylene vessels, the concentration being determined with a den-
simeter. The resultant phases were analyzed by x-ray diffraction.
Individual phases were studied chemically. Analysis of the phases
for sodium was based on a flame photometer of the Carl Zeiss Jena
type, using standard NaCl solutions. The germanium was determined
by titration of its mannite complex with caustic soda [5]. The zinc
was determined complexonometrically [6]. The impurity content
was determined spectrographically.

The experiments were carried out in standard autoclaves
lined with different materials, such as copper, titanium, silver,
N10 steel, and 45KhNMFA steel. The temperature was measured
with chromel—alumel thermocouples. Up to 20 autoclaves were
placed in the furnace at the same time. The horizontal temperature
fluctuation amounted to $\pm 5°C$ and the vertical gradient 5-20°C (on
the outer wall of the autoclave). In individual experiments we used
a furnace with two heating zones so as to be able to regular the
vertical temperature gradient.

At high temperatures the zinc oxide and germanium dioxide
dissolve with the formation of the corresponding aqua complexes.
It was indicated in [7] that on dissolving zinc in concentrated alkali solutions the complex $Zn(OH)_4^{2-}$ was first formed. In dilute
alkali solutions, however, the zinc occurs in the form of ZnO,
$Zn(OH)_2$, $Zn(OH)_3^-$ and $Zn(OH)_4^-$ [8].

The distribution of zinc aqua ions in alkali solutions is shown
in Fig. 1.

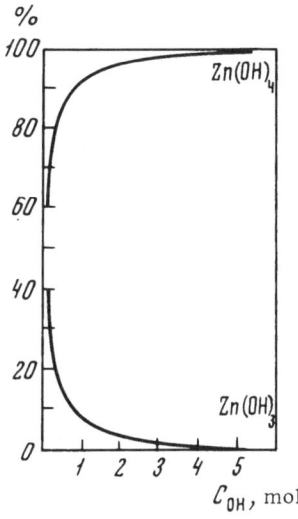

Fig. 1. Dependence of the distribution of zinc aqua ions on the concentration of alkali solutions.

The tendency of zinc toward a tetrahedral coordination is evidently responsible for the existence of aqua complexes of the $[Zn(H_2O)_x(OH)_{4-x}]^{x-2}$ type in aqueous solutions, where the value of x depends on the pH of the medium [9]. In strongly alkaline solutions $Zn(OH)_4^{2-}$ occurs, and here x = 0. On reducing the pH of the solution (to neutral), a dizincate $[Zn_2O(H_2O)_x(OH)_{6-x}]^{x-4}$ may be formed as a result of the association of zincate ions.

The solubility of zinc oxide [4, 10] is independent of pressure and rises with increasing temperature (almost linearly at 200-500°C) and alkali concentration. The solubility of ZnO is 4.5 wt.% in 6.47 M KOH at T = 350°C and P = 550 bar, and 5.7 wt.% in 6.24 M NaOH at T = 350°C and P = 550 bar.

On dissolving in water, germanium dioxide forms germanic acids of various compositions, depending on the pH of the medium [11]. In solution H_2GeO_3 exists in colloidal and molecularly-dispersed states, the dissociation constants being small ($K_1 = 2.6 \cdot 10^{-9}$, $K_2 = 1.9 \cdot 10^{-13}$) [12]. Raising the temperature increases the ionization of H_2GeO_3; this may possibly be due to association with the formation of strong polygermanic acids [13].

The solubility of GeO_2 (at 25°C) rises with increasing pH and reaches a maximum in a 2.5 M solution of NaOH, after which it starts falling. The increase in the solubility of germanium dioxide

in alkali solutions is due to the formation of readily-soluble germanates, possibly with an intermediate stage in the formation of
a surface-hydrated layer [$GeO_2 \cdot aq$], subsequently transforming
into metagermanic acid: $GeO_2 + aq \rightarrow [GeO_2 \cdot aq] \rightarrow H_2GeO_3 + aq.$
In the presence of OH^- ions the reaction $GeO_2 + OH^- \rightarrow HGeO_3^-$
may occur.

The rate of dissolution of GeO_2 in alkalis is determined by
the reaction in question [12, 14].

A study of the $Na_2O-GeO_2-H_2O$ system at 25°C showed [15]
that the solubility isotherm had two minima (at concentrations
of 0.04 and 0.39 mole/liter NaOH), two maxima (at C_{NaOH} =
0.135 mole/liter and C = 2.31 moles/liter NaOH), and a break (at
C = 1.185 moles/liter NaOH). As the alkali concentration increases,
the composition of the residues in equilibrium with the solution
changes thus:

$$Na_2Ge_5O_{11} \cdot 4H_2O \rightarrow Na_2O \cdot 4GeO_2 \cdot 6H_2O \rightarrow Na_2Ge_3O_7 \cdot 7H_2O \rightarrow$$
$$\rightarrow Na_2Ge_2O_5 \cdot 8H_2O \rightarrow Na_2GeO_3 \cdot 7H_2O.$$

The phase diagrams of the Na_2O-GeO_2 and $Na_2O-GeO_2-H_2O$ systems are compared in Fig. 2. The solubility of GeO_2 in 2.31
moles/liter NaOH is 1.163 moles/liter [15]. We may suppose that
with increasing temperature (up to 300-500°C) the solubility of
GeO_2 in water and aqueous solutions of alkalis reaches considerable
values. In alkaline aqueuous solutions germanium evidently forms
aqua complexes of two types:

$$[Ge(H_2O)_x(OH)_{4-x}]^{+x} \text{ and } [Ge(H_2O)_x(OH)_{6-x}]^{x-2},$$

corresponding to coordination numbers 4 and 6 (in contrast to Si,
Ge exhibits both of these in its compounds). Depending on the
concentration of the alkali, the temperature, and the presence of
impurities in solution, the germanate ions may undergo association.
By analogy with silicon [16] it is reasonable to assume that mono-
germanate ions only exist in strongly concentrated alkali solutions.

1.   The $Na_2O-GeO_2-H_2O$ System

A study of this sytem in the temperature range 100-400°C and
at a pressure of 12 atm indicated the existence of two crystalline
products $Na_3HGe_7O_{16} \cdot 4H_2O$ and $Na_2Ge_4O_9$, the latter being determined
as $Na_4Ge_9O_{20}$ [17, 18]. The yield of the first product falls as the
period of the experiment increases, and after 24 h only $Na_4Ge_9O_{20}$

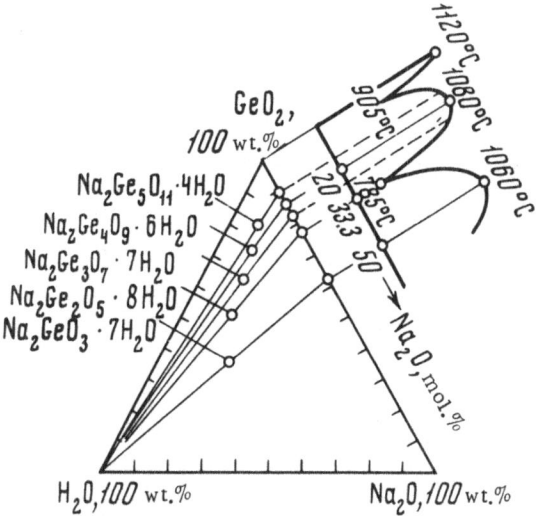

Fig. 2. Phase diagrams of the $Na_2O-GeO_2-H_2O$ and Na_2O-
GeO_2 systems

appears in the reaction products. When a neutral solution of $Ge(OH)_4$
reacts with an aqueous solution of NaOH and these are subsequently
heated in an autoclave at 300°C for 1-6 days, crystals 30-40 μ in
size are obtained [19]. A study of the Na_2O-GeO_2 system revealed
the existence of two congruently-melting compounds, $Na_2O \times GeO_2$
and $2Na_2O \cdot 9GeO_2$, with melting points respectively equal to 1103 ±
15 and 1073 ± 3°C [20]. On rapidly cooling the Na_2O-GeO_2 melt,
sodium tetragermanate crystallizes out; heating this leads to an
exothermic reaction with the formation of $Na_4Ge_9O_{20}$ [21], which
thus constitutes the stable phase of the system in question.

Our own experiments were aimed at expanding hydrothermal
research into a region of higher temperatures and pressures
(500°C and an occupation factor of 75%). Altogether we made 58
experiments with a duration of 7-10 days each; the results are
shown in Table 1. On the basis of these results we plotted the
fields of stability of the phases in T–C coordinates as indicated
in Fig. 3. As the NaOH concentration increased, phases of different
compositions formed in the solution. Changes in temperature and
pressure had no effect on the boundaries of the phases; only re-
duing the temperature to 200°C and under retarded the synthesis
of the germanates. Over the whole range of temperatures studied
with an NaOH concentration of under 3% in the solution (experiments

TABLE 1

Experiment	T, °C	Occupation factor	Concentration of solutions, wt.%	Weight of charge, g	Lining material	Results
1	450	75	1	15	Copper	GeO_2 (tetr.)
2	450	75	2	15	»	The same
3	450	75	3	15	»	»
4	450	75	5	15	Stainless steel	$Na_4Ge_9O_{20}$
5	300	75	5	15	Titanium	The same
6	450	75	10	30	Stainless steel	»
7	300	75	10	30	Titanium	»
8	450	75	10	30	Stainless steel	»
9	200	75	10	20	Titanium	»
10	400	75	20	40	»	»
11	300	75	10	40	»	»
12	480	75	10	40	St. N 10	Phase X
13	450	75	10	40	Copper	$Na_4Ge_9O_{20}$
14	450	75	10	50	Titanium	The same
15	350	75	15	30	St. N 10	»
16	300	75	15	30	Titanium	»
17	300	75	15	30	Copper	»
18	400	75	15	40	»	»
19	450	75	20	40	St N 10	$Na_4Ge_9O_{20}$, phase X
20	300	75	20	40	St.45KhNMFA	Phase X
21	300	75	20	30	St. N 10	$Na_4Ge_9O_{20}$
22	300	70	20	30	Titanium	$Na_4Ge_9O_{20}$, phase X
23	200	80	20	30	»	$Na_4Ge_9O_{20}$
24	400	70	20	40	»	Phase E
25	300	75	20	30	Copper	$Na_4Ge_9O_{20}$
26	400	75	20	40	»	The same
27	300	75	20	60	Titanium	»
28	450	75	20	30	St. 45KhNMFA	$Na_4Ge_9O_{20}$, phase X
29	480	75	20	40	St. N 10	$Na_4Ge_9O_{20}$
30	480	75	20	40	Copper	The same
31	450	75	25	40	St. N 10	Phase X
32	450	75	25	40	St. 45KhNMFA	»
33	480	75	25	40	St. N 10	»
34	300	75	30	40	The same	$Na_4Ge_9O_{20}$, phase N
35	300	75	30	30	St. N 10	Phase X
36	300	75	30	40	Titanium	Phase E
37	300	75	30	40	Copper	$Na_4Ge_9O_{20}$
38	400	75	30	50	»	The same
39	450	75	40	40	Stainless steel	Phase Y
40	300	75	40	40	St. N 10	Phase X
41	300	75	40	40	Titanium	Phase E
42	200	80	40	30	»	GeO_2 (tetr.), $Na_4Ge_9O_{20}$
43	300	75	40	40	Copper	$Na_4Ge_9O_{20}$
44	400	75	40	50	»	The same

Table 1. (Continued)

Experiment	T, °C	Occupation factor	Concentration of solutions, wt.%	Weight of charge, g	Lining material	Results
45	300	70	40	60	Titanium	»
46	450	75	45	60	St. 45KhNMFA	No crystals
47	450	75	60	50	»	Phase N
48	300	80	60	50	St. N 10	Phase X
49	300	75	60	50	Titanium	Phase E
50	200	75	60	40	»	$Na_4Ge_9O_{20}$
51	300	75	60	50	Copper	Phase H
52	300	75	60	60	»	No crystals
53	450	80	80	50	St. 45KhNMFA	Phase Y
54	300	75	80	50	St. N 10	No crystals
55	300	75	80	50	Copper	Phase H
56	400	80	80	60	»	»
57	450	80	80	50	»	»

1-3), the fine-crystalline tetragonal form of germanium dioxide was formed in the "hot" zone of the autoclave. On raising the alkali concentration to 5% (experiments 4-38) crystallization of $Na_4Ge_9O_{20}$ took place, ceasing for NaOH concentrations of over 40%. The $Na_4Ge_9O_{20}$ crystals are formed in the upper and lower parts of the autoclave. In autoclaves containing alkali solutions with a concentration of over 30%, viscous, gelatinous liquids were formed at the end of the experiment, these dissolving completely in water, and only in individual experiments with an alkali concentration of 60-80% did the crystallization of a new phase take place (experiments 55 and 57) (phase H). Chemical analysis of the crystals in question was impossible owing to their small size (tenths of a millimeter) and the difficulty of separating them from the solution in which crystallization occurred.

The $Na_4Ge_9O_{20}$ phase is a colorless, well-faced set of crystals of prismatic habit, 2-3 mm in size, and of tetragonal symmetry, class 4/m (Fig. 4). The lattice constants are a = 14.98 Å, c = 7.38 Å, the calculated density 4.268 g/cm^3, the measured density 4.19 g/cm^3. The cell contains four formula units and is formed by coupled tetrahedra and octahedra. Four octahedra with common faces form the Ge_4O_{16} group. The groups are connected with each other by means of the GeO_4, forming three-dimensional chains

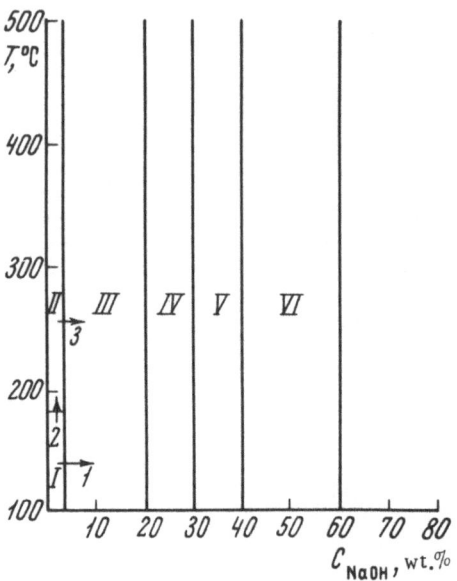

Fig. 3. Ranges of crystallization in the Na_2O-
GeO_2-H_2O system. I) GeO_2 (tetr); II) GeO_2
(hex); III) $Na_4Ge_9O_{20}$; IV) $Na_4Ge_9O_2$, phases E,
X; V) E, X; VI) E, X, N, Y; VII) H, Y.

Fig. 4. Crystals of sodium germanate $Na_4Ge_9O_{20}$ (× 10).

of $(Ge_5O_{16})_n$. By the collectivization of the vertices of the tetrahedra, a $(GeO_3)_n$ spiral is formed; uniting four $(Ge_5O_{16})_n$ chains,
this forms a space lattice [22].

The crystals of the tetragonal form of GeO_2 (tenths to
hundredths of a millimeter in size) are colorless and transparent.
The interplane distances of all three phases obtained in this system
are shown in Table 2. Spectral analysis data are presented in
Table 10. In all the experiments in which the original mixture
(soluble form of germanium dioxide) reacted incompletely with the
NaOH, the hexagonal form of GeO_2 transformed into the rutile form.
As we wished to determine which of the two forms of germanium
dioxide took part in the formation of the germanate, we made some
additional experiments in order to determine the temperature of
the GeO_2 (hex) $\rightarrow$ GeO_2 (tetr.) transition. It is well known that at
temperatures below $1033 \pm 10°$ the stable phase is the tetragonal
form of GeO_2, while the hexagonal form is stable at higher temperatures. The soluble form of GeO_2 formed in certain chemical
reactions (under normal conditions) is evidently metastable and
on heating to 380°C transforms into the rutile form. Certain impurities act as catalysts for this reaction. The use of the hydrothermal method for preparing the insoluble form of GeO_2 showed
that, just as in the case of SiO_2 and the silicates, water catalyzed
the transformation of one form into the other, reducing the temperature of this transformation [23].

TABLE 2.

GeO₂ (tetr.)		Na₄Ge₉O₂₀		Phase II	
I/I_1	d_α/n	I/I_1	d_α/n	I/I_1	d_α/n
100	3.11	99	5.31	38	3.63
80	2.40	100	4.09	51	3.15
20	2.20	84	3.60	100	3.09
17	2.10	50	3.34	53	2.63
5	1.97	25	2.91	10	2.46
70	1.62	35	2.60	10	2.36
20	1.56	40	2.48	14	1.93
		20	2.43	51	1.73
		72	2.37		
		34	2.31		
		36	2.23		

Note: Conditions of recording: anticathode Cu, filter Ni, voltage 25 kV, current 4.5 mA.

Our experiments confirmed that the hexagonal form of GeO_2 transformed into the tetragonal in the presence of H_2O at 185 ± 10°C (f = 75%). In an experiment lasting 48 h, 15 g of the soluble form of germanium dioxide was completely converted into the rutile form. The low solubility of the tetragonal form of GeO_2 is responsible for the difficulty of obtaining GeO_2 crystals and the necessity of growing these at high temperatures (600–700°C) and pressures (~4500 atm) [24].

Thus the dissolution of germanium dioxide and the formation of $Na_4Ge_9O_{20}$ may be pictured in the following way (Fig. 3). The original reagent GeO_2 (hexagonal form) dissolves in aqueous solutions of NaOH with the formation of sodium hydrogermanate $Na_3HGe_7O_{16} \cdot 4H_2O$, subsequently transforming into $Na_4Ge_9O_{20}$ (arrow 1 in Fig. 3). Since the autoclave is being heated, on reaching a specific temperature the reaction GeO_2 (hex) → GeO_2 (tetr) (arrow 2) starts taking place (we remember that the germanium dioxide charge is taken in excess). Subsequent formation of $Na_4Ge_9O_{20}$ takes place as a result of the dissolution of the tetragonal form of GeO_2 (arrow 3) in the presence of a temperature drop in the autoclave and the crystallization of $Na_4Ge_9O_{20}$ in the cold zone of the latter. The formation of $Na_4Ge_9O_{20}$ under hydrothermal conditions takes place by way of an intermediate stage, including the synthesis of the metastable $Na_3HGe_7O_{16} \cdot 4H_2O$. The existence of the latter is confined to a brief time interval [17], and it failed to appear perceptibly in our experiments on the system under consideration. The formation of potassium hydrogermanate and the hydrogermanates of other metals will be considered later.

In order to protect them from corrosion by the alkali solutions, the autoclaves were lined with a variety of materials. The "purest" experiments were carried out in silver and copper linings. Titanium and steels of the N10 and 45KhNMFA types corroded to a certain extent and reacted with the dissolved germanates, leading to the formation of "side" reaction products. The individuality of the phases obtained was established by x-ray diffraction (Table 3). In certain cases chemical and spectral analyses were carried out (Table 10). For convenience the phases are indicated by letters.

TABLE 3

Phase E — Na TiO [GeO$_4$]		Phase X		Phase N		Phase Y	
I/I_1	d_α/n	I/I_1	d_α/n	I/I_1	d_α/n	I/I_1	d_α/n
90	5.207	41	4.50	60	3.72	20	4.28
10	4.12	52	3.26	100	3.14	22	2.76
11	3.50	80	3.07	20	2.63	100	2.56
22	3.36	100	3.01	10	2.36	18	2.32
100	2.81	90	2.54	45	2.00	15	1.95
30	2.53			10	1.77	30	1.55
21	2.41			8	1.67		
40	2.27			10	1.51		
40	1.74						
33	1.67						
30	1.53						
30	1.49						

Note: Conditions of recording: anticathode Cu, filter Ni, voltage 25 kV, current 4.5 mA.

Fig. 5. Crystals of phase E, Na$_2$(TiO)[GeO$_4$] (× 8).

Ph a s e E, composition $Na_2 (TiO) [GeO_4]$ (experiments 24, 36, 41, 49, Table 1). Crystals of phase E are only formed in titanium linings with an alkali concentration of over 20%. The crystals are of lamellar habit (Fig. 5) and the basic forms are $\{001\}$, $\{201\}$, $\{111\}$. The crystallographic system is tetragonal. The crystals are transparent, colorless, and 3–4 mm in size. With increasing synthesis temperature and alkali concentration the crystals become larger.

Ph a s e X, composition $NaFeGe_2O_6$, is formed in linings made from either of the types of steel mentioned (experiments 12, 19, Table 1). The crystals are yellow, transparent, and of elongated prismatic form, 1–1.5 mm in size. This phase is formed for an alkali concentration of 10% or over, independently of the temperature and pressure. The results of a spectral analysis are presented in Table 10.

An x-ray structural analysis of phase X yielded the monoclinic lattice parameters and confirmed the isostructural relationship between $NaFeGe_2O_6$ and $NaMgSi_2O_6$ [25].

Ph a s e N is formed under the same conditions as phase X; it is represented by colorless, well-faced crystals of prismatic habit, 4–5 mm in size, which appear in the charge at the bottom of the autoclave in small quantities (experiments 34, 47, Table 1). The chemical composition remains unknown.

TABLE 4

Experi- ment	Weight of charge, g		Ratio ZnO/GeO_2	Results
	ZnO	GeO₂		
1	10	10	1 : 1	Phase V
2	5	15	1 : 3	"
3	15	5	3 : 1	Phase V, ZnO
4	10	10	1 : 1	Phase V, GeO_2 (tetr.)
5	5	15	1 : 3	Phase V
6	5	15	1 : 3	"

Phase Y. The conditions of formation are the same as in the previous two cases (alkali concentrtion 40-80%). The formation of this phase is independent of temperature and pressure. The crystals are formed in the "hot" zone of the autoclave in quite large quantities; they have a tetragonal habit and are of a green color, with conical depressions on the faces, being 1-3 mm in size (experiments 39, 53, Table 1). On exposure to air the crystals are covered with a black deposit. The spectral analysis of the crystals is presented in Table 10.

2. Crystallization in the ZnO−GeO$_2$−H$_2$O System

Investigations were carried out at 450°C with an autoclave occupation factor of 75% (the autoclaves were lined with copper) and weight ratios of the oxides in the charge equal to 1:1, 1:3, and 3:1. The results of six experiments are presented in Table 4, from which we see that the stable phase is in this case the germanium analog Zn$_2$GeO$_4$ of the natural compound willemite Zn$_2$SiO$_4$, and only for an excess of ZnO in the charge does zincite crystallize.

Phase V, Zn$_2$GeO$_4$ (Fig. 6), is of the trigonal type and crystallizes in the form of hexagonal prisms $\{11\bar{2}0\}$ together with a rhombohedron $\{12\bar{3}1\}$. The crystals are colorless and transparent,

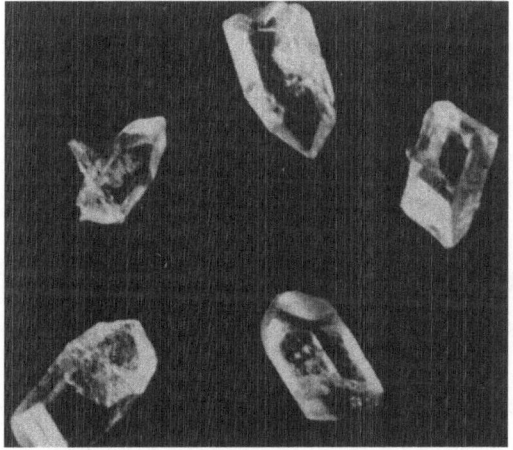

Fig. 6. Crystals of willemite, Zn$_2$GeO$_4$ (× 2).

TABLE 5

Experiment	T, °C	f, %	Concentration of solutions, wt.%	Weight of charge, g	Lining material	Results
1	400	75	1	1	Copper	GeO_2 (tetr.)
2	300	75	1	30	St. 45KhNMFA	The same
3	300	75	2	30	St. N 10	»
4	450	75	3	1	Copper	$K_2Ge_4O_9$
5	300	75	3	20	St. N 10	The same
6	300	75	3	30	St. 45KhNMFA	$K_2Ge_4O_9$, GeO_2 (tetr.)
7	200	75	5	15	St. N 10	No crystals
8	300	75	5	30	The same	$K_2Ge_4O_9$, GeO_2 (tetr.)
9	450	75	5	30	»	The same
10	450	80	5	15	Titanium	»
11	300	75	5	20	St. N 10	$K_2Ge_4O_9$, phase K, GeO_2 (tetr.).
12	350	75	5	15	The same	$K_2Ge_4O_9$
13	350	75	5	30	»	$K_2Ge_4O_9$, GeO_2 (tetr.),
14	350	75	5	30	Copper	$K_2Ge_4O_9$
15	450	75	5	15	St. N 10	The same
16	400	75	5	20	Titanium	»
17	450	75	5	15	Copper	Phase K, $K_2Ge_4O_9$
18	450	75	5	1	»	The same
19	300	80	5	25	St. 45KhNMFA	»
20	300	75	5	30	St. N 10	»
21	300	75	10	30	The same	$K_2Ge_4O_9$, GeO_2 (tetr.)
22	450	80	10	15	»	The same
23	200	80	10	15	»	No crystals
24	450	75	10	15	St. N 10	$K_2Ge_4O_9$, GeO_2 (tetr.), phase K.
25	450	75	10	15	Titanium	$K_2Ge_4O_9$, GeO_2 (tetr.)
26	450	75	10	3	»	$K_2Ge_4O_9$, phase K, GeO_2 (tetr.)
27	450	75	10	30	St. N 10	$K_2Ge_4O_9$
28	350	75	10	30	The same	The same
29	350	75	10	30	Copper	Phase K
30	400	75	10	15	Titanium	$K_2Ge_4O_9$, phase K
31	450	75	10	30	Copper	The same
32	450	75	10	30	»	$K_2Ge_4O_9$, phase K
33	450	75	15	20	Iron	$K_2Ge_4O_9$
34	400	75	15	15	Titanium	The same
35	300	75	15	40	Copper	Phase K
36	350	70	40	40	St. 45KhNMFA	$K_2Ge_4O_9$, phase K
37	450	75	15	40	Stainless steel	The same
38	450	75	20	15	The same	$K_2Ge_4O_9$

Table 5. (Continued)

Experiment	T, °C	t, %	Concentration of solutions, wt%	Weight of charge, g	Lining material	Results
39	450	75	20	20	Titanium	$K_2Ge_4O_9$
40	450	75	20	40	Stainless steel	»
41	450	75	20	20	The same	$K_2Ge_4O_9$, phase K
42	400	75	20	15	Titanium	$K_2Ge_4O_9$
43	350	75	20	40	St. 45KhNMFA	$K_2Ge_4O_9$ phase K
44	450	75	20	40	Copper	The same
45	400	75	·25	20	Titanium	$K_2Ge_4O_9$
46	350	75	25	40	St. 45KhNMFA	The same
47	450	75	25	40	Stainless steel	$K_2Ge_4O_9$, phase R
48	450	75	30	50	St. N 10	$K_2Ge_4O_9$
49	400	75	30	20	Titanium	$K_2Ge_4O_9$, phase K
50	350	75	30	50	St. 45KhNMFA	$K_2Ge_4O_9$, phase K
51	450	75	30	40	Copper	The same
52	450	75	35	60	St. N 10	$K_2Ge_4O_9$
53	400	75	40	20	Titanium	Phase K
54	300	75	40	50	St. 45KhNMFA	No crystals
55	450	75	40	50	Stainless steel	Phase R
56	450	70	40	50	Copper	Phase K
57	400	75	40	50	»	»
58	450	80	45	40	Stainless steel	Phases R, K
59	450	75	45	50	The same	Phase R
60	300	75	45	50	St. 45KhNMFA	No crystals
61	400	75	50	50	St. N 10	Phase R
62	400	75	50	25	Titanium	Phase M
63	300	80	50	50	St. 45KhNMFA	Phase R
64	450	75	50	70	Copper	$K_2Ge_4O_9$, phase K
65	300	60	50	80	»	Phase K
66	400	75	60	50	St. N 10	Phase R
67	450	75	60	50	Copper	No crystals
68	400	75	60	50	»	Phase K
69	450	80	60	70	»	$K_2Ge_4O_9$, phase K
70	300	60	65	80	»	Phase K
71	450	60	80	80	»	»
72	450	75	80	50	»	No crystals
73	300	60	80	80	»	$K_2Ge_4O_9$, phase K
74	450	75	80	70	»	$K_2Ge_4O_9$
75	400	75	80	60	»	The same
76	400	75	80	50	»	»
77	300	80	80	60	»	»

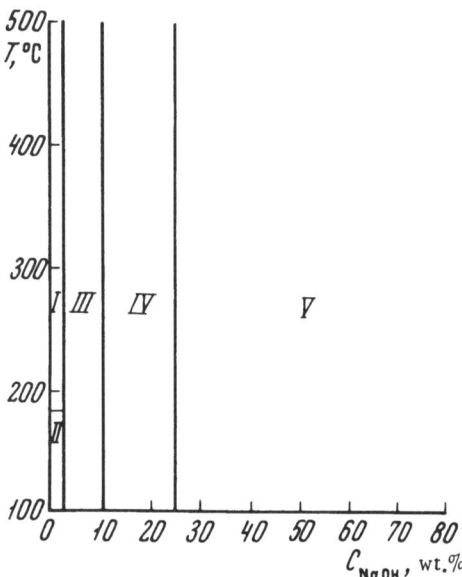

Fig. 7. Regions of crystallization in the K_2O-GeO_2- H_2O system. I) GeO_2(tetr); II) GeO_2 (hex); III) GeO_2 (hex); $K_2Ge_4O_9$, K; IV) $K_2Ge_4O_9$, K, M; V) $K_2Ge_4O_9$, K, R, M.

or tinted orange by iron. Crystals activated with manganese display bright green tribo-, cathodo-, x-ray-, and photoluminescence. The maximum of the emission spectrum is at 537 nm on excitation by ultraviolet light at λ = 365 nm.

3. Crystallization in the $K_2O-GeO_2-H_2O$ System

A study of crystallization from a mixture of KOH + GeO_2 in the presence of water under conditions identical with those governing the formation of $Na_4Ge_9O_{20}$ revealed the existence of a crystalline product with a composition of $K_2HGe_7O_{16} \cdot 4H_2O$ [17]. No other germanates of the same type as the sodium germanates were found. Crystallization in the K_2O-GeO_2 system revealed two crystalline phases of composition $K_2Ge_4O_9$, $K_4Ge_9O_{20}$, isotypic with the corresponding sodium compounds. Potassium tetragermanate is also obtained by roasting the zeolite $K_3HGe_7O_{16} \cdot 4H_2O$ at over 700°C. Heating the zeolite above 600°C leads to the formation of $K_4Ge_9O_{20}$ (with a strongly exothermic reaction); on further heating, potassium tetragermanate is formed, this being the stable phase of the system in question, in contradistinction to the stable $Na_4Ge_9O_{20}$ in the Na_2O-GeO_2 system [21].

The aim of our investigations lay in studying the $K_2O-GeO_2-H_2O$ system at high temperatures (300-450°C) and pressures (f = 60-75%). Each experiment lasted 7-10 days under steady conditions. The KOH concentration in the solution was varied between 1 and 80 wt.%. The results are presented in Table 5. The stability fields of the phases so formed are indicated in Fig. 7. The boundaries of the crystallization fields of the various phases are determined solely by the alkali concentration and are independent of temperature and pressure. For a caustic potash concentration of 1-2% in the solution, the tetragonal form of germanium dioxide is formed (experiments 1-3). In the concentration range 2-10% KOH, crystals of tetragonal germanium dioxide, potassium tetragermanate, and crystals of composition $K_3HGe_7O_{16} \cdot 4H_2O$, provisionally called phase K, develop (experiments 4-47). The individual nature of the resultant phases may be established optically and by x-ray diffraction. The formation of the potassium tetragermanate and $K_3HGe_7O_{16} \cdot 4H_2O$ was followed up to a concentration of 80% KOH in the solution. For larger alkali concentrations viscous, gelatinous liquids (soluble in water) were frequently observed. X-ray data for the resultant phases are presented in Table 6. Spectral-analysis data appear in Table 10.

TABLE 6

$K_2Ge_4O_9$		Phase K		$K_2Ge_4O_9$		Phase K	
I/I_1	d_α/n	I/I_1	d_α/n	I/I_1	d_α/n	I/I_1	d_α/n
9	3.79	27	4.43	3	2.27	20	2.05
23	3.45	70	3.84	3	2.24	20	1.92
9	3.16	6	3.43	3	2.20	21	1.86
8	3.04	100	3.14	3	2.12	30	1.81
27	2.97	7	2.85	3	1.97	7	1.76
100	2.83	7	2.79	3	1.85	30	1.72
62	2.81	75	2.71	23	1.83	13	1.57
15	2.54	40	2.56	45	1.71	27	1.54
5	2.50	80	2.43	10	1.65		
30	2.46	50	2.32	30	1.46		
30	2.38	16	2.22				

Note: Conditions of recording: anticathode Cu, filter Ni, voltage 25 kV current 4.5 mA.

Fig. 8. Crystals of potassium germanate, $K_2Ge_4O_9$
(× 8).

The crystals of the tetragonal form of germanium dioxide
have already been described.

The $K_2Ge_4O_9$ crystals are of prismatic habit, colorless, short-
columnar or elongated along the c axis, sometimes lamellar, of
hexagonal symmetry, and 1-8 mm in size (Fig. 8). In the majority
of the experiments these crystallize in the "cold" zone of the autoclave.

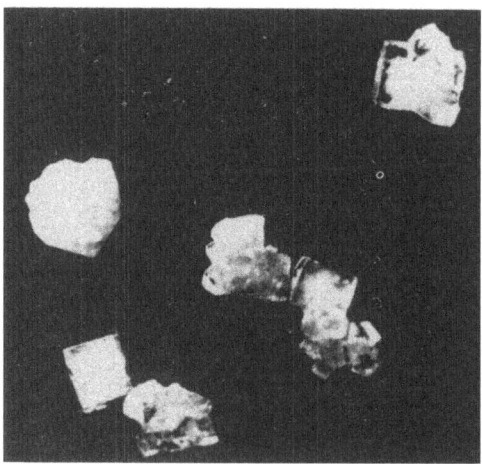

Fig. 9. Crystals of potassium hydrogermanate
$K_3HGe_7O_{16} \cdot 4H_2O$ (× 10).

Crystals of the $K_3HGe_7O_{16} \cdot 4H_2O$ phase are of cubic habit, colorless, transparent or cloudy as a result of gas and liquid inclusions, and 1-3 mm in size (Fig. 9). Germanates of composition $Me_3HGe_7O_{16} \times 4H_2O$ are obtained under normal conditions for all the alkali metals and also for $(NH_4)^+$, Ag and Tl [2], all these being isostructural. Although the radii of the ions vary considerably, the lattice parameter remains almost constant. The structure is characterized by a combination of two coordination polyhedra: octahedra (GeO_6) and tetrahedra (GeO_4) (Fig. 10). The three-dimensional chains of the octahedral groups are joined by way of the GeO_4 tetrahedra, as in the structure of $Na_4Ge_9O_{20}$ [22], except for the fact that in the present case they may be distinguished in all three directions. Large channels are formed in these in the [100], [010] and [001] directions in which the ions of the alkali metals are situated. The zeolitic properties of these crystals may be attributed to this kind of structure. It should be noted that, in the simultaneous presence of Na and K, K and Mn, or K and Ti in the solution, the crystallization of the corresponding germanate takes place more intensively. Chemical-analysis data relating to potassium hydrogermanate also agree closely with the calculated concentrations of its constituent elements; in the simultaneous presence of K^+ and Na^+ in the solution, a "mixed" hydrogermanate $(K_2Na) \cdot HGe_7O_{16} \cdot 4H_2O$ is formed.

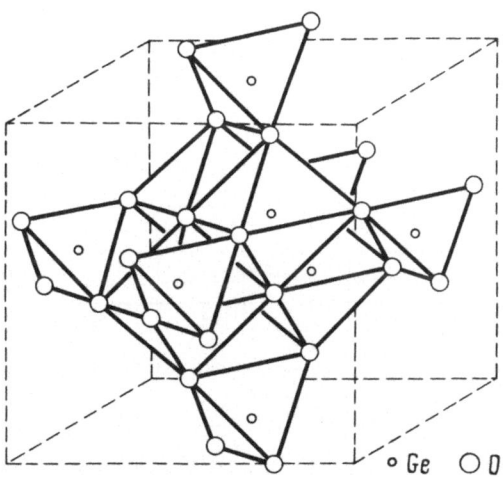

Fig. 10. Structure of $K_3HGe_7O_{16} \cdot 4H_2O$.

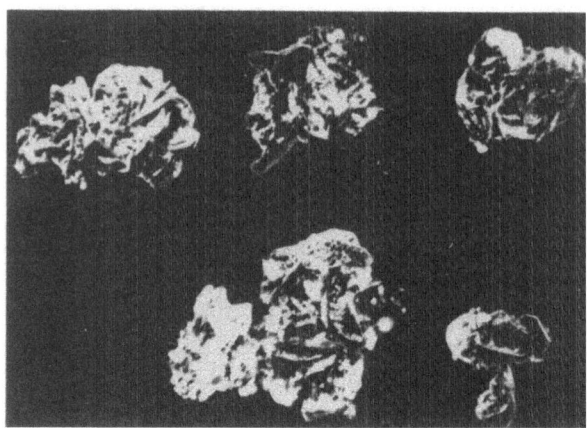

Fig. 11. Crystal concentrations of the phase R(×30).

The alkali corrosion of the linings of the autoclave used for the crystallization led to the formation of a number of "side" phases; an x-ray study of these gave a number of sharp and readily-indexed x-ray diffraction patterns.

Phase R is represented by scaly laminar crystals of light-brown color and mica-like habit growing together in the form of "roses"; it is formed for an alkali concentration of 25% or over (Fig. 11).

Phase M appears in the form of fibrous crystals 1-5 mm in size, being formed in titanium-lined autoclaves. The interplane distances of these phases are given in Table 7, and the spectral analysis in Table 10.

TABLE 7

Phase R		Phase M		Phase R		Phase M	
I/I_1	d_α/n	I/I_1	d_α/n	I/I_1	d_α/n	I/I_1	d_α/n
100	3.46	40	2.98	5	1.59	70	2.11
47	2.61	40	2.75			40	2.08
15	2.09	40	2.71			25	1.58
7	1.74	100	2.59				

Note: Conditions of recording: anticathode Cu, filter Ni, voltage 25 kV, current 4.5 mA.

4. Crystallization in the
$K_2O - ZnO - GeO_2 - H_2O$ System

We were the first to study this system. The experiments
were carried out at 300-450°C with an autoclave occupation factor
of 75% in 1-80% solutions of potassium hydroxide. Altogether 40
experiments were carried out; the results appear in Table 8. The
crystallization fields of the various phases are shown in Fig. 12.
As in all the other systems which we have just been describing,
phase formation depends on the alkali concentration and not the
temperature and pressure. We see from Fig. 12 that over a wide
range of KOH concentrations (1-40%) the stable phase is germanium
willemite Zn_2GeO_4, phase V. For an alkali concentration of over
40% no crystals were found to form, and only for a large excess
of zinc oxide in the charge did the crystallization of zincite occur
(experiment 28). On further increasing the KOH concentration in
solution a new phase KD crystallized; the composition of this was
not determined owing to the difficulty of separating it from the
solution in which crystallization occurred (experiments 35-40,
Table 8). The individuality of the phase was determined by x-ray
diffraction.

Phase KD is represented by colorless tabular crystals 1-3
mm in size having a hexagonal habit and often forming concretions.
The interplane distances are given in Table 9 and spectral-analysis
data in Table 10.

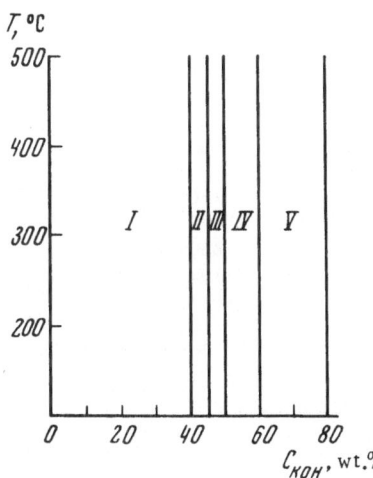

Fig. 12. Regions of crystallization in the
$K_2O-ZnO-GeO_2-H_2O$ system. 1) Phase V
Zn_2GeO_4; II) phases V and R; III) phase R;
IV) no crystals formed; V) phase KD.

TABLE 8

Exper-iment	T, °C	Concen-tration of solutions, wt.%	Weight of charge, g		Lining material	Results
			ZnO	GeO$_2$		
1	300	1	15	15	Copper	Phase V
2	450	1	15	15	St. N 10	»
3	450	3	15	15	»	K$_2$Ge$_4$O$_9$, phase V
4	350	5	15	15	»	Phase V
5	350	5	15	15	Copper	Phase V
6	450	5	15	15	St. N 10	»
7	450	5	15	15	Copper	»
8	400	5	15	15	St. N 10	K$_2$Ge$_4$O$_9$, phase V
9	375	5	15	15	Copper	»
10	350	10	15	15	St. N 10	Phase V
11	350	10	15	15	Copper	»
12	450	10	15	15	St. N 10	»
13	450	10	15	15	Copper	»
14	450	20	15	15	Stainless steel	»
15	450	20	15	15	Copper	»
16	300	25	15	15	St. 45KhNMFA	»
17	450	30	15	15	Stainless steel	»
18	450	35	15	15	St. N 10	»
19	450	30	30	30	Copper	»
20	300	30	30	30	»	»
21	300	40	15	15	St. 45KhNMFA	Phase R
22	450	40	15	15	Copper	ZnO
23	450	40	30	30	»	Phase V
24	400	40	30	30	»	Phases V, R, ZnO
25	300	40	30	30	Copper	Phase V
26	450	45	40	40	Silver	Phases V, ZnO
27	400	50	15	15	St. N 10	Phase R
28	300	50	50	40	Copper	ZnO
29	400	60	15	15	St. N 10	No crystals
30	300	60	15	15	St. 45KhNMFA	»
31	450	60	15	15	Copper	»
32	450	80	15	15	Stainless steel	Phase KD
33	450	60	30	30	Copper	»
34	400	60	30	30	»	ZnO
35	300	60	30	30	»	Phase KD
36	450	70	40	40	»	ZnO, phase KD
37	450	80	30	30	»	Phase KD
38	400	80	30	30	»	The same
39	300	80	30	30	»	»
40	450	80	40	40	»	»

TABLE 9

I/I_1	d_α/n	I/I_1	d_α/n
48	4.92	100	2.84
100	3.20	42	2.63
51	3.07	28	2.53
100	3.02	27	2.39
31	2.92	30	2.25

Note: Conditions of recording: anticathode Cu, filter Ni, voltage 25 kV, current 4.5 mA.

As a side phase in the concentration range 40-50% KOH we obtained phase R, which was described in Section 3.

Conclusions

The reactions taking place between the germanate and zincate ions present in solution and the solvent lead to the formation of poorly-soluble sodium and potassium germanates and zincogermanates of the $R_x Zn_y Ge_p O_q$ type, where R = Na or K. Hydro-

TABLE 10

Impurity	Phase D	Phase K	Phase H	Phase Y	Phase X	Phase KD	$K_2 Ge_4 O_9$
Mn	0.006	0.001	0.07	0.1	0.1	0.07	—
Al	0.005	0.02	0.005	0.0005	0.035	—	0.005
Ca	0.035	0.035	0.05	0.005	0.05	0.035	0.02
Mg	0.003	0.002	0.005	0.005	0.005	0.005	0.005
Cu	0.0005	0.368	0.363	0.005	0.035	0.05	—
Na	$\geqslant 0.5$	$\geqslant 0.5$	$\geqslant 0.5$	$\geqslant 0.5$	0.5	—	—
Ag	0.0005	—	—	—	0.005	—	—
Si	>0.5	0.2	0.0275	0.5	$\geqslant 0.5$	—	0.005
Zn	$\geqslant 0.5$	0.005	$\geqslant 0.5$	—	0.005	$\geqslant 0.5$	—
Fe	0.5	0.36	>0.5	$\geqslant 0.5$	$\geqslant 0.5$	0.36	0.01
Ge	>0.5	>0.5	>0.5	>0.5	>0.5	>0.5	>0.5
Bi	—	0.035	—	—	—	—	—
V	—	0.0275	—	—	0.2	—	—
K	—	>0.5	–	—	—	>0.5	>0.5
Cr	—	–	—	>0.5	>0.5	0.005	0.005
Ni	—	—	—	0.2	0.5	—	—

Note: Impurity content given in %.

thermal conditions promote the occurrence of these reactions. A study of the structure of the germanates shows that the differences in the coordinations of Ge and Si manifest themselves in the properties of the resultant compounds: all the silicates of the alkali metals so formed are readily soluble in water, while most of the germanates dissolve sparingly. Well-formed germanate crystals may therefore be formed under hydrothermal conditions.

It is also important to note the formation of a poorly-soluble form of GeO_2 of the rutile type.

1. We have shown that it is fundamentally possible to synthesize sodium and potassium germanates and zincogermanates under hydrothermal conditions.

2. As in the case of silicate systems, so also in the germanate systems phase formation depends on the alkali concentration; within the ranges studied, temperature and pressure have no effect on the formation of phases.

3. In the $Na_2O-GeO_2-H_2O$ system the stable phase is $Na_4Ge_9O_{20}$, which is formed over a wide range of NaOH concentrations. No metastable $Na_3HGe_7O_{16} \cdot 4H_2O$ appears to be formed.

4. In the $K_2O-GeO_2-H_2O$ system the stable phase is $K_2Ge_4O_9$ and the metastable phase $K_3HGe_7O_{16} \cdot 4H_2O$; however, the latter is formed over almost the whole range of C_{KOH}, and is evidently more stable than sodium hydrogermanate. Impurities certainly have a stabilizing influence on $K_3HGe_7O_{16} \cdot 4H_2O$.

5. In the $K_2O-ZnO-GeO_2-H_2O$ and $ZnO-GeO_2-H_2O$ systems, germanium willemite (Zn_2GeO_4) is formed over almost the whole of the range studied.

6. The simultaneous presence of zincate and germanate ions in solution leads to the formation of new Zn-Ge compounds.

7. We have confirmed that the GeO_2 (hex) $\rightarrow GeO_2$ (tetr) transformation takes place at $185 \pm 10°C$ under hydrothermal conditions.

Literature Cited

1. I. P. Kuz'mina, O. K. Mel'nikov, and B. N. Litvin, in: Hydrothermal Synthesis of Crystals (A. N. Lobachev, ed.) [in Russian], Izd. Nauka (1968), p. 141.
2. H. Nowotny, Uspekhi Khim., 27:997 (1958).

3.   B. N. Litvin and O. K. Mel'nikov, Kristallografiya, 14:101 (1969).

4.   R. A. Laudise, E. D. Kolb, and A. J. Caporaso, J. Amer. Ceram. Soc., 47:9 (1964).

5.   V. F. Gillebrand, R. É. Lendal', R. A. Brait, and D. I. Gofman, Practical Hand-
     book on Inorganic Analysis [in Russian], Goskhimizdat (1960), p. 320.

6.   Yu. N. Knipovich and Yu. V. Morachevskii, Analysis of Raw Minerals [in Rus-
     sian], Goskhimizdat (1959), p. 481.

7.   T. P. Dirkse, J. Electrochem. Soc., 101:328 (1954).

8.   I. I. Treadford, J. Amer. Chem. Soc., 76:6022 (1954).

9.   N. V. Belov, Miner. Sb., No. 15, Note 72 (1961).

10. R. A. Laudise, Amer. Mineral., 48:642 (1963).

11. H., Brintzinger, Z. Anorg. Chem., 256:98 (1948).

12. E. A. Knyazev, I. A. Kakovskii, and Yu. B. Kholmanskikh, Zh. Neorg. Khim.,
     10:2698 (1965).

13. V. A. Nazarenko and A. M. Andrianov, Uspekhi Khim., 34:8 (1965).

14. V. A. Vekhov, B. S. Vitukhnovskaya, and R. F. Doronkina, Zh. Neorg. Khim.,
     11:237 (1966).

15. E. A. Knyazev and S. V. Borisova, Zh. Neorg. Khim., 12:2785 (1967).

16. R. K. Ailer, Colloidal Chemistry of Silica and Silicates [in Russian], Goss-
     troiizdat (1959).

17. E. R. Shaw, J. F. Corwin, and J. W. Edwards, J. Amer. Chem. Soc., 80:1536
     (1958).

18. I. F. White, E. R. Shaw, J. W. Edwards, and A. Pabst, Analyt. Chem. 31:315
     (1959).

19. N. Ingri and G. Kundgren, Arkiv. Kemi, 18:479 (1961).

20. M. K. Murthy and J. Aquavo, J. Amer. Ceram. Soc., 47:444 (1964).

21. A. Wittman and P. Papamantellos, Monatsh. Chem., 91:855 (1960).

22. N. Ingri and G. Lundgren, Acta Chem. Scand., 17:617 (1963).

23. A. N. Laubengauer and D. S. Morton, J. Amer. Chem. Soc., 54:2303 (1932).

24. M. L. Harwill and R. Roy, Amer. Ceram. Soc. Bull., 44:297 (1965).

25. L. P. Solov'eva and V. V. Bakalin, Kristallografiya, 12:591 (1967).

Crystallization Kinetics of Sodium Zincogermanate Na_2ZnGeO_4

I. P. Kuz'mina, A. N. Lobachev, and N. S. Triodina

Sodium germanates and zincogermanates have recently been synthesized in the $Na_2O-ZnO-GeO_2-H_2O$ system at high temperatures and pressures and converted into the form of crystals with a variety of compositions. Some of these compounds had never been detected before, for example, the crystalline phase of composition Na_2ZnGeO_4, provisionally named phase D [1]. Crystals of this phase are formed at 200-250°C at pressures corresponding to autoclave occupation factors of 50-80% in sodium hydroxide solutions with concentrations of 5 wt.% or over. In the concentration range 5-15 wt.% NaOH, the phase crystallizes together with certain other crystalline phases; for concentrations of over 15 wt.%, only the crystals of phase D are formed over the whole temperature and pressure range studied. The original charge employed for the synthesis incorporated the chemical reagents zinc oxide (chemically pure and especially pure), and germanium dioxide (especially pure). Dissolving in aqueous solutions of caustic soda, the zinc oxide and germanium dioxide react with each other and with the solvent, the final product being a poorly-soluble compound of composition $Na_2O \cdot ZnO \cdot GeO_2$, formed by the following reaction:

$$ZnO + GeO_2 + 2NaOH \rightleftarrows Na_2ZnGeO_4 + H_2O.$$

The crystals are mainly formed in the "hot" zone of the autoclave. In the presence of a temperature drop, transport occurs and the compound (phase D) crystallizes in the upper ("cold") part of the autoclave. The spontaneously-formed crystals of phase D grow to dimensions of 2-8 mm, they are transparent and colorless. Crystals containing traces of Fe^{+3} have an orange tint.

The structure of the compound in question was studied by N. V. Belov et al., who found that it belonged to the monoclinic system and had lattice constants of $a = 8.91$ Å, $b = 5.60$ Å, and $c = 5.33$ Å; $\beta = 127°$. The Fedorov group was Pa, and the unit cell contained two formula units [2]. The interplane distances are given in Table 1.

Preliminary investigations showed that the crystals were characterized by strong intracrystalline fields and exhibited a piezoelectric effect. Crystals activated with manganese luminesce under photo-, x-ray, and cathodo-excitation; they exhibit tribo- and electroluminescence.

Preliminary measurements were made on spontaneously-formed crystals of phase D. Subsequently samples oriented in particular crystallographic directions had to be selected. To this end a method of growing phase D single crystals on a seed was developed.

Growth took place for temperatures of 340-400°C in the dissolution zone and 280-320°C in the growth zone, the temperature drop being 20-80° (along the outer wall of the autoclave) and the pressure 400-1500 atm, in solutions of 30 wt.% NaOH.

In order to determine the true temperatures and temperature drops, the temperature distribution inside the autoclave was mea-

TABLE 1. Interplane Distances of
Na_2ZnGeO_4 (λFe)

I	d_α/n	I	d_α/n	I	d_α/n
6	4.40	1	2.23	1	1.57
5	4.28	1	2.02	4	1.53
1	3.39	4	1.99	2	1.507
6	3.02	1	1.95	2	1.488
8	2.79	1	1.81	2	1.450
6	2.66	2	1.79	1	1.413
10	2.62	1	1.72	2	1.402
2	2.47	1	1.69	1	1.348
1	2.33	6	1.65	1	1.315
				1	1.313

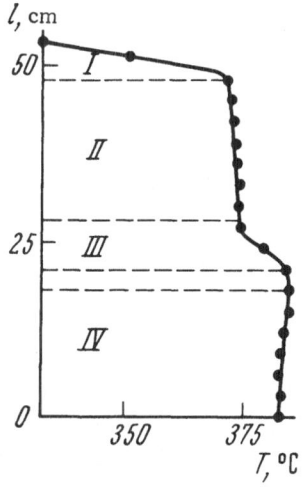

Fig. 1. Temperature distribution inside the autoclave. I) Closure; II) growth zone; III) diaphragm; IV) charge.

sured in a special autoclave using the same equipment as that used for growing the crystals and the same solvents. The results were compared with the readings of the thermocouples on the outside of the working autoclave (Fig. 1). The values of ΔT_{ext} °C subsequently obtained in working experiments could therefore readily be converted to the true temperature drops within the autoclave.

The crystals were grown in silver vessels of two types: a "floating" vessel of volume ~ 850 cm^2 and a lining-type vessel of volume ~ 80 cm^3.

In all cases the zones of dissolution and growth were separated by a silver or Teflon disc constituting a barrier, with apertures having a total area of 3-5% of the area of the inner cross section of the lining. The ratio of the volume of the growth and dissolution zones was varied from 1 : 1 to 3 : 1 in particular experiments. At the top of the inner vessel was a silvered copper cover and a metal stopper.

The "floating" vessel was placed in the autoclave in such a way as to give a gap of 10-20 mm between the sides of the autoclave and inner vessel; into this, water or alkali solutions of a concentration lower than that in the vessel itself were poured. It was then required to created approximately equal pressures inside and outside the inner vessel so as to create a hermetic state. The autoclaves were heated in individual furnaces of the resistance type with two heating zones; this facilitated the creation of the desired temperature drop. In the autoclaves with a "floating" inner vessel the pressure was established and maintained by means of a contact manometer regulating the heating of the autoclave. The experiments lasted 5-25 days in the steady condition. The seeds employed were spontaneously-formed crystals and plates cut from Na_2ZnGeO_4 crystals in a specific direction.

The spontaneously-formed crystals and the crystals grown on seeds were bounded by four dihedra (011), (01$\bar{1}$), (110), ($\bar{1}$10) and four monohedra (001), (00$\bar{1}$), (100), ($\bar{1}$00). Experiments showed that the morphological shape of the crystals depended on the crystallization temperature. Thus at T = 260-350°C all the foregoing forms were present in the crystals. The most developed faces were the (110), ($\bar{1}$10) and (001) (00$\bar{1}$) and the least developed were the (100), ($\bar{1}$00), (011), (01$\bar{1}$). Under these conditions the growth rates v of all the faces of the crystal were commensurable. Increasing the temperature to 350°C or over changes the morphology of the crystals. Together with a general increase in the growth rates of all the faces, the growth rates of the faces of the two dihedra (011) and (01$\bar{1}$) increase particularly sharply, and this leads to a rapid tapering effect. Crystals obtained under these conditions exhibit the following simple forms: four monohedra (001), (00$\bar{1}$), (100), ($\bar{1}$00) and only two dihedra (110) and ($\bar{1}$10), the other two dihedra being absent.

Crystals formed at temperatures of under 250°C are bounded by four dihedra (011), (01$\bar{1}$), (110), ($\bar{1}$10) and two monohedra (100) and ($\bar{1}$00). A temperature fall of this nature reduces the growth rate of all the faces but has the least effect on the growth rates of the monohedra (001) and (00$\bar{1}$), so that these grow relatively swiftly and tapering occurs.

We see from all the foregoing arguments that the most important growing faces of the phase D crystals are the faces of the monohedra (001), (00$\bar{1}$) and the dihedra (011) and (01$\bar{1}$).

We made a quantitative measurement of the growth rates of the different faces expressed in mm/day of grown layer. At temperatures of 250 and 300°C and with various temperature drops we measured the growth rates of the faces (00$\bar{1}$), (001), (01$\bar{1}$), (011). The results are presented in Table 2. Experiment established the following sequence of growth rates v for the faces of phase D crystals at different temperatures:

for T < 250°C

$$v_{(001)} > v_{(00\bar{1})} > v_{(0)} > v_{(100)} > v_{(110)};$$
$$v_{(001)}/v_{(011)} > 1;$$

for T = 250–350°C

$$v_{(011)} \geqslant v_{(001)} > v_{(00\bar{1})} > v_{(100)} > v_{(1\bar{1}0)};$$
$$v_{(001)}/v_{(011)} \leqslant 1;$$

for T ≥ 350°C

$$v_{(011)} > v_{(001)} > v_{(00\bar{1})} > v_{(100)} > v_{(110)};$$
$$v_{(001)}/v_{(011)} < 1.$$

The growth rates of the (001) and (00$\bar{1}$) faces and the growth rates of the faces of the dihedra (011) and (01$\bar{1}$) are commensurable and differ by 15–20%.

A graphical relationship between the growth rates of the various faces for a crystallization-zone temperature of 300°C and various temperature drops ΔT is illustrated in Fig. 2. We see from this figure that the growth rate of all the faces increases linearly with increasing temperature drop, $v_{(011)}$ increasing more sharply than $v_{(001)}$ and $v_{(00\bar{1})}$. On extrapolating the straight lines we see that they intersect the horizontal axis for temperature drops of ΔT ≈ 8-10°C, which probably constitute the critical supersaturations (ΔT_{cr}) for all three faces. For constant temperature and autoclave occupation factor, and for materials with a direct relationship between solubility and temperature, such as phase D, the supersaturation increases linearly with the temperature drop between the dissolution and growth zones. We may therefore study the ef-

TABLE 2. Growth Rates of Different Faces of Na_2ZnGeO_4 Crystals

Serial No.	T, °C	ΔT_{ext} °C	v, mm/day			
			(00$\bar{1}$)	(001)	(011)	(01$\bar{1}$)
1	300	25	0.15	0.182	0.22	—
2	250	45	0.19	0.22	0.30	—
3	300	40	0.30	0.35	0.42	—
4	300	65	0.60	0.63	0.75	0.76
5	300	50	0.33	0.44	0.54	0.40
6	300	60	0.50	0.58	0.65	0.70

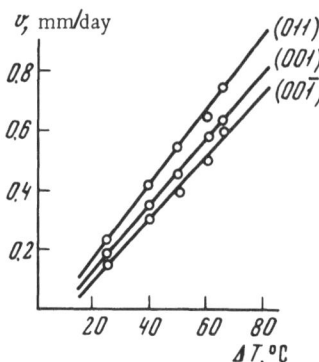

Fig. 2. Relation between the growth
rates of the faces of phase D crystals
and the temperature drop for a crystallization-zone temperature of 300°C.

fect of the relative supersaturation on the kinetic growth characteristics of Na_2ZnGeO_4 by simply varying the temperature drop.

The resultant critical supersaturations for the (001), (00$\bar{1}$),
and (011) faces differ by no more than 1-2%, in excellent agreement with the fact that no sharp anisotropy appears in the growth
rates of the different faces, which in a number of experiments
justifies the use of spontaneously-formed phase D crystals as seeds.

The influence of pressure on the crystallization rate of the
phase was only estimated qualitatively in our experiments: The
growth rates of the faces of sodium zincogermanate crystals increase with increasing autoclave occupation factor and diminish
when the latter is reduced.

A change in the ratio of the oxides in the charge ZnO/GeO_2
has hardly any effect on the growth rates of the phase D crystals.

By studying the effects of various factors on the crystallization process we were able to determine the optimum conditions
for the growth of the crystals: $T = 300 \pm 10°C$, $\Delta T = 30\text{-}40°C$
(along the outer wall of the autoclave), $P = 400\text{-}600$ atm, $C_{NaOH} =$
30 ± 5 wt.%. The seed was oriented parallel to the (001) or (011).

In experiments lasting 14 days we obtained crystals several
cm^3 in volume and up to 40 g in weight. By increasing the time
of the experiment, still larger crystals may be obtained.

The sodium zincogermanate crystals so grown are often
imperfect; they contain cracks and gas and liquid inclusions, and
exhibit hazing. Low temperatures (250°C and under) result in the
formation of a large number of gas-liquid inclusions within the

crystals, this evidently being associated with the increasing viscosity of the medium and the high probability of its being captured
during growth.

For temperatures of over 350°C, crystals with a large
number of cracks are often formed, probably owing to the high
growth rate at these temperatures. Medium temperatures of 250–
350°C are therefore the most acceptable.

On rapidly cooling the autoclave, the temperatures of the growth
and dissolution zones swiftly become equalized, and the autoclave
cools with almost identical temperatures in both zones; the
already-grown phase D crystals are then etched by the alkali
solutions, and the surfaces of the faces become mat, with traces
of dissolution.

The formation and growth of the crystals under hydrothermal conditions is not accessible to observation, but it may nevertheless be studied by examining the microtopography of the crystal surfaces, which reflects the phenomena taking place on the
face. The necessity of producing more perfect crystals led to a
more intensive examination of the growth process. A macro- and
micromorphological study of the relief on the faces of the crystals
was accordingly carried out. The geometrical characteristics of
the faces were determined by means of an MIM-8M metallographical
microscope. The effect of the relative saturation (ΔT) on the morphology of the surfaces of various faces was also considered.

The faces of the dihedron (110) are characterized by a layer-
like growth, the layers propagating in the direction of the z axis
(Fig. 3).

The relief on the faces of the other dihedron (011) is slightly
different. This face exhibits rapid growth. Owing to the intensive
capture of particles from the solution by active growth centers,
new layers develop on layers which are still in the incomplete
stage, and so on. The surface becomes characterized by uneven,
disoriented hillocks or projections (Fig. 4). One section of the
same face exhibited growth accessories in the form of conical
hillocks (Fig. 5). The face of the dihedron (011) exhibits conical
growth accessories of another form also (Fig. 6). On the sloping
sides we see groups of quasi-parallel layers, partly intersecting
each other and creating patterns. These may well be due to either

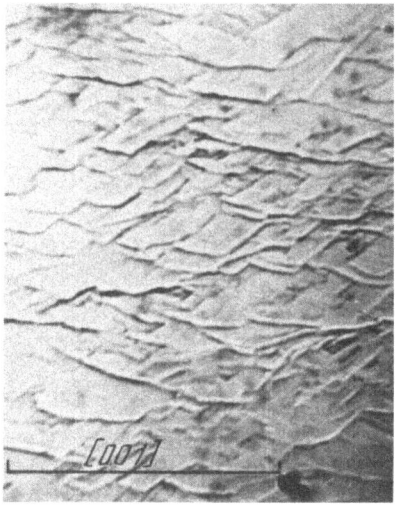

Fig. 3. Layer-like growth on the face
of the dihedron (110) (× 320).

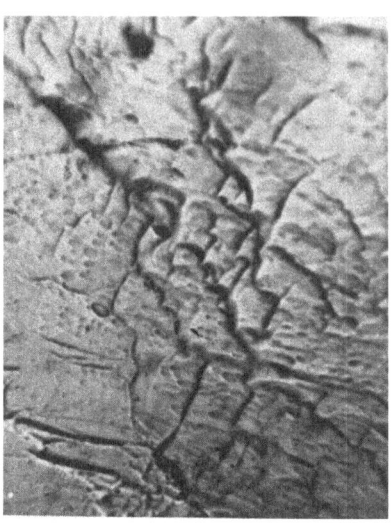

Fig. 4. Surface of the rapidly-growing
face of the dihedron (011) (× 160).

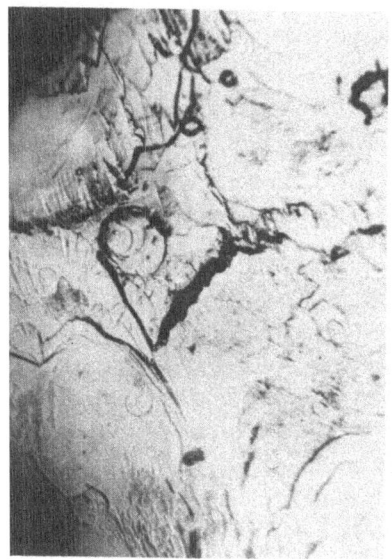

Fig. 5. Growth surface of the face of
the dihedron (011) (× 80).

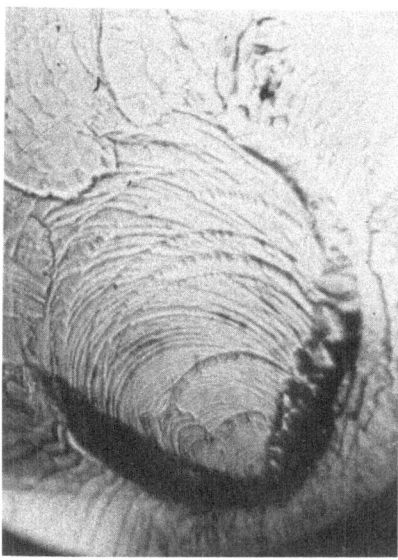

Fig. 6. Conical layer-like growth figures on the (011) face (× 160).

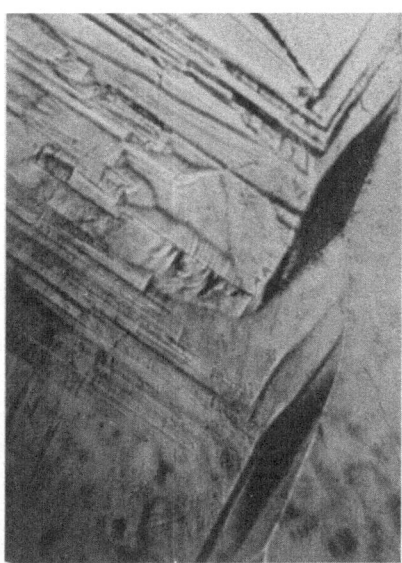

Fig. 7. Growth figures on the (001) face (× 80).

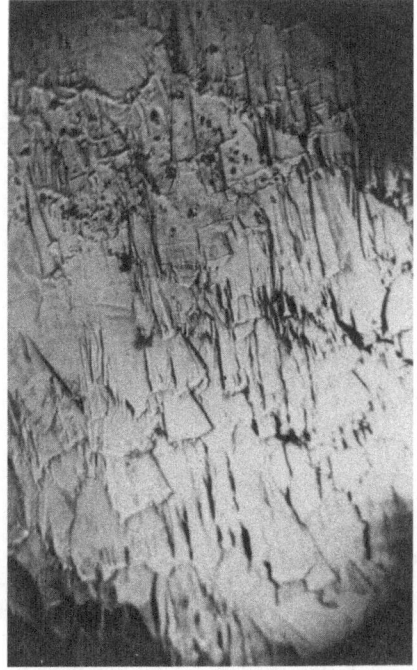

Fig. 8. Trapezoidal plane growth figures on the (001) face (× 80).

Fig. 9. Growth figures in the form of tetrahedral pyramids transforming into cones (× 80).

Fig. 10. Surface of the (001) face (× 80).

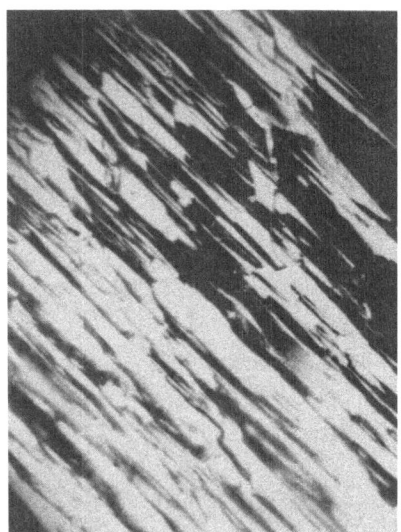

Fig. 11. Wedge-shaped growth figures on the (001) face (× 80).

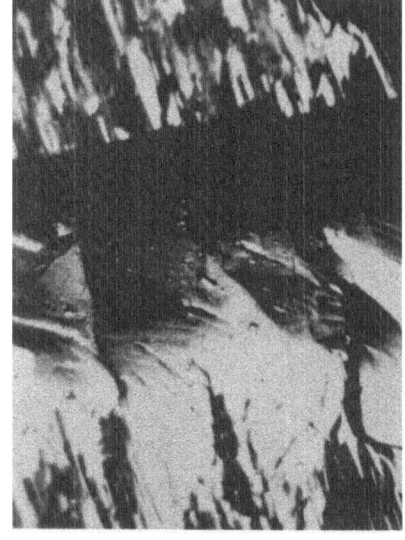

Fig. 12. Rounded growth figures on the (001) face (× 80).

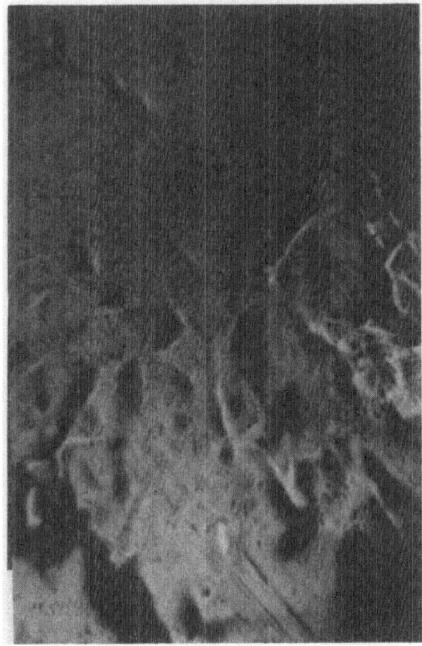

Fig. 13. Surface of the face of the dihedron (110) (× 80) after etching in HCl
solution.

Fig. 14. Etch pattern on the surface of the (110) dihedral face due to a 30%
solution of NaOH (× 160). a) Figures in the form of horizontal streak projections; b) figures in the form of truncated pyramids.

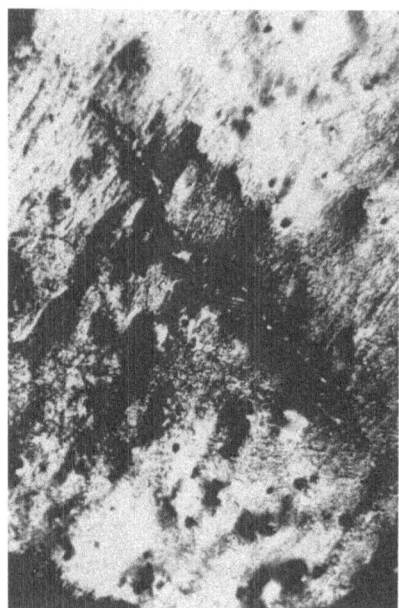

Fig. 15. Etched face of the dihedron
(011) (× 80).

Fig. 16. Etch pattern of phase D crystals
grown in the presence of an oxidizing agent
(KMnO$_4$) (× 80).

a large colloidal particle or a very small crystallite, subsequently covered over with newer and newer layers.

On the face of the monohedron (001) we observe regions of two different types: One is covered with angular stepped projections (Fig. 7), and the other by trapezoidal plane growth figures (Fig. 8). The face of the monohedron (001) is least of all subject to dissolution, and it exhibits growth figures in the form of tetrahedral pyramids, transforming into cones, the vertices of these being sharply inclined in the direction of the [010] axis (Fig. 9). Observing the change in the relief of the rapidly-growing (001) faces on increasing the relative supersaturation ΔT in the solution during growth, we perceive that the surface of the face acquires a different nature: The growth layers become thicker and cover the whole face, forming a unilateral stepped structure in the form of projections, possibly arising from a single vertex (Fig. 10).

The $(00\bar{1})$ face of the monohedron is covered with uniform figures (Fig. 11) alternating with regions not affected by dissolution, with rounded growth figures (Fig. 12).

We also studied the faces of the crystals after dissolution in acid and alkali solutions and observed several morphologically different types of etch figures.

The etch patterns of the (110) face in HCl solutions are shown in Fig. 13.

On etching with alkali solutions (30 wt.% NaOH), horizontal streak projections appear on the faces of the dihedra (110) (Fig. 14a), as well as figures in the form of truncated pyramids, drawn out in the [001] direction (Fig. 14b).

The surface of the (011) face of the dihedron exhibits no particular sharp relief; its relief is expressed in the form of smoothed dissolution figures, and is very reminiscent of the pattern observed on crystals treated in HCl (Fig. 15).

On studying crystals obtained under the same conditions but in the presence of an oxidizing agent ($KMnO_4$) we observed a characteristic confinement of the etch pits to the vertices of the growth accessories, the formation of which was probably associated with dislocations (Fig. 16).

The foregoing growth and dissolution figures are characteristic of the majority of phase D crystals grown under hydrothermal conditions.

Conclusions

1. By studying the crystallization processes of Na_2ZnGeO_4
we have been able to deduce the general laws governing the growth
of the crystals and determine the optimun conditions (T, ΔT, P,
C) for producing sodium zincogermanate crystals up to 40 g in
weight.

2. We have studied the macro- and microrelief of the faces
of the crystals grown in this way. We have determined the details
of the relief characterizing the various faces and established their
relationship to the conditions of crystal growth.

Literature Cited

1.   I. P. Kuz'mina, O. K. Mel'nikov, and B. N. Litvin, in: Hydrothermal Synthesis
      of Crystals [in Russian], Izd. Nauka (1968), p. 141.
2.   É. A. Kuz'min, Author's abstract of Dissertation [in Russian], IKAN SSSR,
      Moscow, (1968).

Controlling the Growth of Crystals in Autoclaves

A. A. Shternberg

The main difficulty which arises in mastering the growth of crystals under hydrothermal conditions lies in the fact that there is little information available as to the processes taking place in the autoclave, and it is accordingly difficult to envisage ways on controlling them.

In an earlier communication [1] we set out all available objective data unambiguously related to the growth of crystals in an autoclave; analysis of these data enabled us to make some well-founded changes in subsequent experiments and also to reproduce earlier results in autoclaves of different shapes and greater sizes.

We are well aware that the growth of crystals by the temperature-drop method involves the motion of the solid material from the "hot" part of the autoclave to the "cold" part. This displacement of the material in the autoclave may be observed during the experiment as a displacement of mass, of the center of gravity of the system, or of the volume of the solid phase.

Thus specially-equipped systems should enable us to follow the kinetics of the recrystallization process and hence control it over the period of the experiment by varying, for example, the temperature difference between the crystal growth and dissolution zones.

Controlled Growth of Crystals
in a Horizontal Autoclave

The growth of the crystals may be followed very closely by
observing the displacement of the mass in a horizontal autoclave.
Figure 1 shows the fundamental arrangement of a "weight" apparatus in which the autoclave, as it were, plays the part of the
arm of an analytical balance. The horizontal autoclave 1 is placed
in the furnace 2 without touching it, its center of gravity resting
on a vertical axle 3 made of stainless steel. The lower end of the
axle passes outside the furnace and the safe 4 in which it is en-
closed, and stands on the triangular prism 5, which allows the system to oscillate freely. To prevent the movable system from tilting,

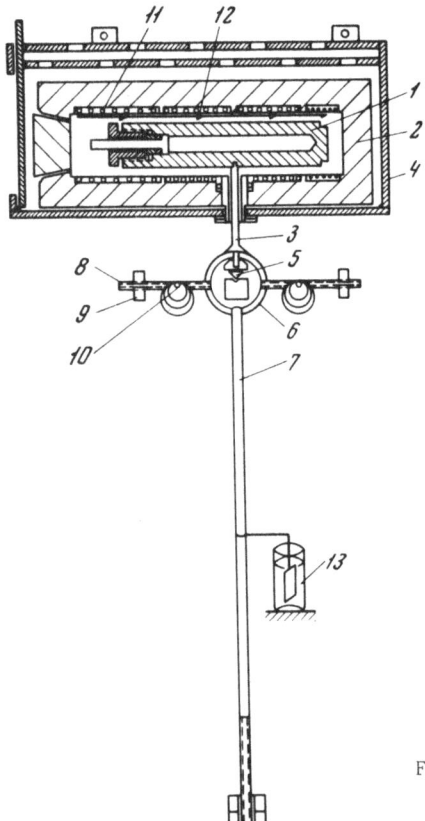

Fig. 1. Arrangement of balance system.

it is connected to a counterweight 7 by means of the ring 6, the weights being selected in such a way as to make the center of gravity of the system as a whole almost coincide with the point of support. Two horizontal axles 8 pass into the ring 6; the weights 9 moving along these serve to adjust the system into the zero position before the start of the experiment. On each of the axes 8 (immediately underneath the zones respectively accommodating the charge and the seed in the autoclave) we place cross bars 10 designed to hold annular sets of weights; these may be moved from cross bar to cross bar in order to compensate the transfer of mass which takes place in the autoclave during the experiment.

The furnace 2 has four independently-adjustable, sectioned heaters 11. Two of these heat the zones in which the charge dissolves and the crystals grow, and the other two serve to eliminate temperature gradients within each of these zones independently. The temperature in each of the zones in the furnace is monitored by means of the thermocouples 12. These are placed in the furnace space without touching the autoclave. In order to determine the true temperature distribution in the growth zones, a steel, platinum-shielded jacket (not shown in the figure) is introduced into the autoclave to accommodate a sliding thermocouple, which is read periodically.

Since the mass only passes from one end of the autoclave to the other during the growth of the crystals, the system may be brought into equilibrium by moving weights in the reverse direction from one cross bar to the other. The amount of recrystallizing material will be greater than the mass of the weights moved, since the mass transfer is partly compensated by the weight of the displaced liquid. The mass of the grown crystals (m) is given by the relation

$$\frac{m}{m_1} = \frac{\rho}{\rho - \rho_1},$$

where m_1 is the mass of the weights moved, ρ is the density of the crystals, and ρ_1 is the density of the medium in the autoclave.

The position of the reading mark on the counterweight 7 is read through a telescope with a magnification of 100, so that a mass movement of 0.05 g may be detected. In order to prevent spontaneous rocking of the system, it is supplied with an oil brake 13. Of course we cannot know at exactly what point the charge is being dissolved, and the absolute accuracy of the "weighing" is thus no better than 10-20%; nevertheless it is sufficient to indicate what is happening in the autoclave.

Thus on carrying out experiments in the balence system we obtain continuous information regarding the motion of mass in the autoclave, i.e., we know when homogenization of the medium sets in, whether recrystallization of the material is taking place, and at what velocity the recrystallization is proceeding. If crystals are being grown on seeds with a prespecified area, we shall quite easily be able to calculate the rate of growth of the crystals and to regulate this by varying the temperature difference between the dissolution and growth zones.

The horizontal position of the autoclave, together with the recording of the mass transfer taking place in it, provides us with far more information as to the properties of the solution—crystal system under consideration than could be obtained in ordinary experiments, and all in a single process.

1. Since the horizontal autoclave is symmetrical with respect to gravitational forces, the growth and dissolution zones differ only in temperature and may be interchanged with each other. Recrystallization of the material may therefore be carried out in either direction, and in a single experiment we may synthesize the material, recrystallize it in order to average out the composition, and then, in order to produce reasonably large crystals, carry out a third recrystallization in the opposite direction.

2. Since we know how much material is placed in the autoclave and how many grams of it are transferred from one side to the other during the recrystallization, we may estimate the solubility of the material from the weight loss.

3. Considerably more accurate data may be obtained (in the same experiment) as to the temperature coefficient of the solubility of the material. For this purpose, after all the charge has been recrystallized into one end of the autoclave, without changing the sign of the gradient was may vary the temperature in the furnace and note the mass transfer associated with dissolution or crystallization. Here we must remember that in crystallization or dissolution the material passes out into the whole volume, or is concentrated from the whole volume into one of the ends, rather

than passing strictly from one end of the autoclave to the other. Hence in using Eq. (1) for calculating the solubility we must double the result.

On using the foregoing method to obtain the temperature coefficient of solubility, we obtain the solubility isochore of the material, which is limited in the low-temperature direction by the degree of homogenization of the medium in the autoclave, and in the high-temperature direction by the limiting mechanical strength of the latter. We must immediately point out that, when growing crystals in an autoclave by the temperature-drop method, the isobaric solubility of the material is "operative"; in the low-pressure region (up to 700-1000 atm) this has a small inclination to the temperature axis, the sign of which may indeed reverse. The solubility isochores may be effectively used for growing crystals by the method of reducing the temperature, which as yet has never been used in autoclaves.

4. In a vertical autoclave, concentration segregation (phase separation) of the medium often occurs owing to the high solubility of the material. This is less likely with a horizontal autoclave, but even if it occurs it cannot prevent the transfer of material and the growth of the crystals.

5. Independently of whether a direct or retrograde solubility occurs under the conditions chosen, mass transfer will occur, and only its direction may vary; we shall know of this at once from the displacement of the mass in the autoclave.

As examples of the use of the balance apparatus, let us consider some of the results of our own experiments.

Spontaneous Growth of
Sphalerite Crystals

The arrangement of the experiment was as follows. The platinum-lined autoclave had a volume of 137.5 ml, the charge was a finely-divided powder of zinc sulfide of the type used for phosphors (50 g in weight), and the density of the sphalerite equaled 4.0. The true volume of the charge taken was 50/4 = 12.5 ml. The free volume of the autoclave was 137.5-12.5 = 125 ml, and the solvent was 100 ml of orthophosphoric acid with a specific gravity of 1.7 (autoclave occupation factor 80%). The density of the medium in the autoclave during the experiment was 170/125 = 1.36. The sensitivity of the installation in the experiment was 9 scale divisions to 1 g of movable weights; from Eq. (1) we thus have

$$m = m_1 \frac{4}{4 - 1.36} \approx 1.5m_1.$$

Hence the recrystallization of 1 g of sphalerite caused the reading mark to be displaced by six divisions of the microscope scale.

The program of the experiment envisaged the following:

1. Obtaining data as to the solubility of sphalerite in concentrated orthophosphoric acid at 475°C,
2. Plotting the solubility isochores of sphalerite,
3. Producing crystals suitable for further investigations.

The experiment lasted 15 days and comprised the following stages.

1. In order to concentrate the charge originally spread all over the autoclave into one of the ends, a temperature of 480°C was established in the left-hand section of the furnace and 455°C in the right-hand section. The furnace and autoclave were heated over a period of 3 h and the reading mark moved through 78 divisions (from 100 to 178), owing to the expansion of the liquid and the recrystallization of the charge; after this mass transfer in the autoclave ceased.

2. After 17 h, recrystallization of the material was started in the opposite direction by setting a temperature of 450°C in the left-hand section of the furnace and 500°C in the right. The recrystallization lasted for 43 h and ceased at the mark 11. The reading had moved through 167 divisions, so that in this period 167/6 = 28 g of sphalerite had been moved. Thus under the conditions in question the solubility of ZnS was 50-28 = 22 g in 100 ml of solvent.

3. After this, in order to determine the temperature coefficient of the solubility of the material, we reduced the temperature of the whole autoclave by 20°C (to 480 and 430°C respectively). Over a period of 8 h the reading changed by 12 divisions, which corresponded to the crystallization of (12/6) · 2 = 4 g of material, or 0.2 g/deg for 100 ml of solution.

4. Further reductions of 25° in the temperature (to 465 and 415°) and holding for 16 h led to the separation of 4.8 g of material (or 0.19 g/deg in 100 ml of solvent in the lower temperature range).

5. Then the temperature was raised to the original value (450 and 500°C) and held for 5.5 h. Some 6.7 g of material passed into solution in this time. We should have expected that 4 + 4.8 = 8.8 g would have dissolved. Clearly a period of 5.5 h was insufficient to establish equilibrium (saturation of the solution), although no movement of the balance arm had appeared in the last 1.5 h. It might possibly be that the disagreement between the results obtained in the forward and reverse directions may have been due to a side phenomenon, viz., the corrosion of the autoclave through pores (cracks) in the platinum lining. The fact that corrosion had been occurring was plain from the yellow color of the resultant crystals.

6. After plotting the sphalerite solubility data, the temperature gradient was again reversed and crystal growth started, being carried out in two stages.

During the growth of the crystals, the temperature gradient was reduced to zero by means of the additional heater in the left-hand section (in the present case in the growth zone) in order to spread them along the autoclave. We should therefore expect that the mass-transfer effect recorded by the pointer readings would diminish by between 1/4 and 1/3, depending on precisely where the crystals were in fact formed.

In order to grow the crystals by recrystallization we established the following temperatures: left-hand side 485, right-hand side 463°C. The transfer of the material proceeded for 190 h and then ceased. During this time the reading changed by 107 divisions, i.e., $(107/6) \cdot (4/3) = 24$ g of material were transferred.

Then growth was continued at the expense of material still in the dissolved state by gradually reducing the temperature of the autoclave over a period of 48 h. At temperatures of 330 and 369°C (in the left- and right-hand sections) an intensive vibration of the system began, indicating the appearance of a gas phase in the autoclave, and the experiment was stopped. On reducing the temperature, on average, from 478 to 320°, the reading changed by 41 divisions, i.e., another $(41/3) \cdot (4/3) = 18$ g of material were deposited.

Opening the autoclave confirmed that crystallization had occurred in a zone 10 cm long; together with a large number of fine aggregated crystals, single individuals up to 12 mm across had been formed (Fig. 2). The crystals had a yellow coloring, which indicated the corrosion of the autoclave walls. The isometric development of the larger crystals and their small degree of twinning suggested that they had grown principally in the final stage of the process, when the temperature was being reduced.

Recrystallization of Sodalite

In an experiment carried out in conjunction with O. K. Mel'nikov we determined the isochoric temperature coefficient of solubility of synthetic hydrosodalite in a 30% solution of NaOH, which in the temperature range 380-420°C equaled 0.15 g/deg in 100 ml (autoclave occupation factor 75%). The whole charge (20 g) was recrystallized, and a second, slow recrystallization (lasting 13 days) was carried out, as a result of which extremely perfect, well-faced crystals up to 5 mm across were obtained (Fig. 3). In the recrystallization and growth of the crystals, carried out at an average temperature of 400°C, some 11 g of sodalite were transferred from one end of the autoclave to the other; the rest (9 g) thus

Fig. 2. Sphalerite crystal (× 7).

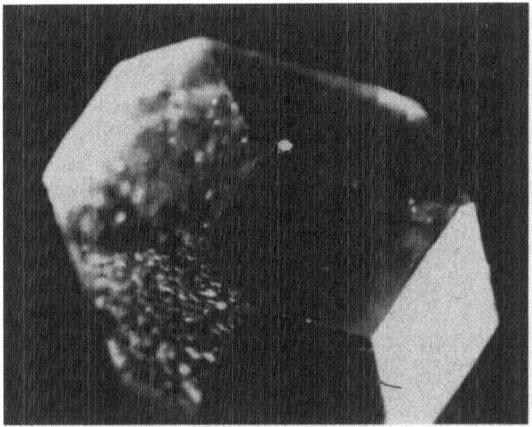

Fig. 3. Sodalite crystal (× 10).

remained in solution. Hence the solubility of sodalite at 400°C in a 30% NaOH solution equals 9 g per 100 ml of solution.

Solubility Isochore of Potassium Niobate

In one of our experiments on growing potassium niobate crystals, we determined the isochoric solubility in a 52% solution of caustic potash at 500-600°C, with 75% occupation of the autoclave. The results are depicted in Fig. 4. The flattening of the curve as temperature falls clearly shows why hardly any recrystallization of potassium niobate is observed below 500°C [2].

These experiments suggest that a balance device of the kind here described may give considerable aid in choosing effective solvents for a particular compound, such as will provide efficient

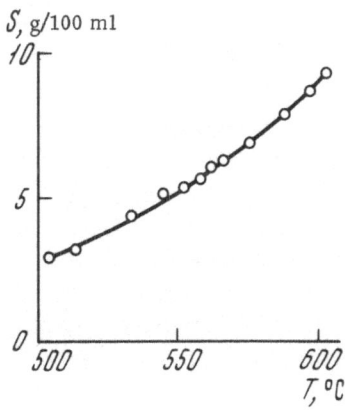

Fig. 4. Solubility isochore of potassium niobate $KNbO_3$ in 52% caustic soda solution.

mass transfer, in optimizing the thermal conditions of the process, and in growing large crystals suitable for investigation, i.e., it may prove useful at all stages of laboratory work aimed at producing single crystals.

For carrying out tests under conditions closer to industrial, we may use the following means of following the flow of the process in a vertical autoclave.

Pendulum Apparatus for Observing the Kinetics of Recrystallization in an Autoclave

Mass transfer may be determined continuously or periodical-ly as the motion of the center of gravity of a vertical autoclave mounted in a device of the kind illustrated schematically in Fig. 5.

The autoclave 1, with a jacket for a sliding thermocouple 2, is furnished with a barrier containing the tubes 3, separating the growth and dissolution zones of the crystals. In the dissolution zone is a container with the charge 4; in the growth zone is a frame holding the crystal seeds 5. The autoclave is placed in a cylindrical descending furnace 6 with two independently adjustable electric heaters 7 and two thermocouples 8. The furnace and autoclave carry noncontiguous sections of a thermally-insulating bellows-type barrier 9, stopping the convective flow of heat in the furnace. The autoclave is connected by means of the rigid axle 10 to the counterweight 11, which is similar in shape and mass to the auto-clave. In its middle section the axle 10 has a point of support, viz., a window with a trihedral prism or a roller bearing on a stationary bracket 12. The counterweight 11 has two balance shelves 13, placed symmetrically along the oscillation axis relative to the centers of the crystal growth zone (lower shelf) and the dissolution zone (upper shelf).

Continuous pendulum oscillations of a harmonic nature are created by some suitable method in the moving system, the devia-tion from the mean position being through a small angle only. (By experience with balance devices, we may expect that oscillations will develop spontaneously as a result of the movement of the liquid in the autoclave and the air around it.) The oscillations are recorded and counted by means of a photo-relay and pulse counter. Knowing the number of oscillations executed over a reasonably

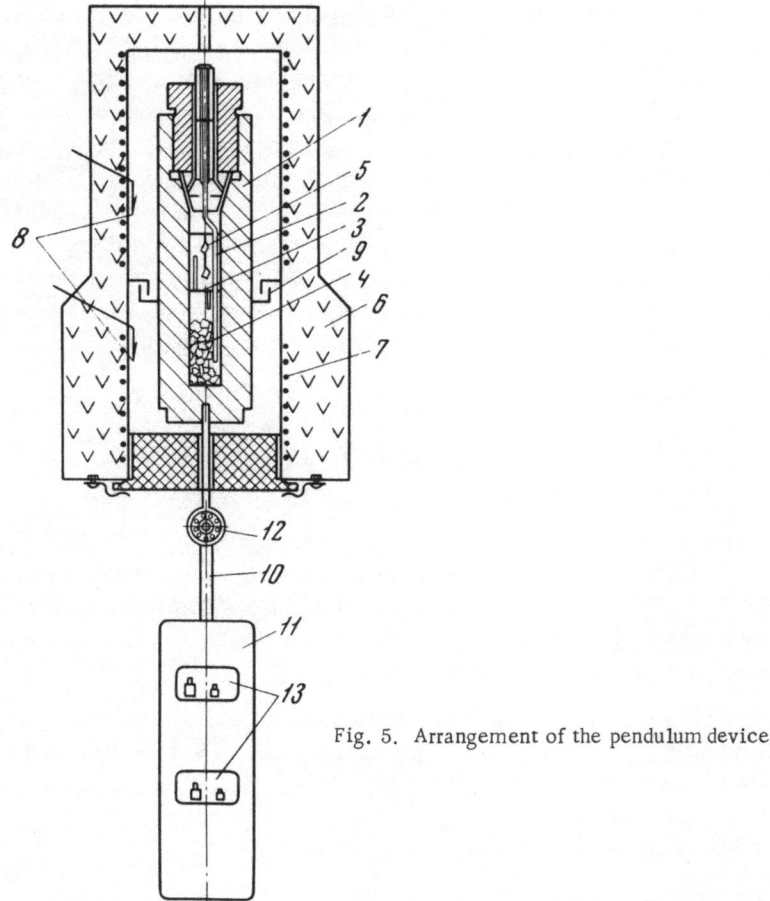

Fig. 5. Arrangement of the pendulum device.

long period of time (hours or days), we may calculate the period of oscillation of the system to a high accuracy.

Crystal growth is followed by observing the change in the oscillation period of the system. If the crystals are actually growing in any experiment, the center of gravity of the upper arm (the autoclave) will move upward and the oscillation period of the system will increase. For an equal motion of the mass, the oscillation period will increase the more, the closer the center of gravity of the system originally lay to the oscillation axis.

Quantitative measurements of the kinetics of the process may be carried out in two ways.

1. Weights approximately equal to the weight of the autoclave charge are placed on the upper shelf on the counterweight, the auto- clave temperatures are adjusted to conform with the crystal-growth mode, and the "zero" oscillation period is then measured; as the crystals grow, the oscillation period will increase. By moving weights from the upper shelf to the lower we shall shorten the oscillation period to its original value, and from the mass of the moved weights we may estimate the amount of crystalline material which has grown on the seeds. Clearly the amount of material transferred in the autoclave wil exceed the mass of the weights so moved by an amount equal to the mass of the solution displaced by the crystal.

The true weight of the transferred material will be

$$m = m_1 \frac{\rho_{cr}}{\rho_{cr} - \rho_{av}},$$

where m_1 is the mass of the weights moved, ρ_{cr} is the specific gravity of the crystal, ρ_{av} is the specific gravity of the medium in the autoclave (weight of the solution poured in divided by the volume of the autoclave after subtracting the volume of the structures in- troduced, the charge, and the seeds).

Knowing the weight of the transferred material, its density, and the volume and area of the seeds placed in the autoclave, we may easily calculate the growth rate of the crystals over a specified period of time and regulate this by varying the temperature dif- ference between the growth and dissolution zones.

We must remember that on changing the thermal conditions in the autoclave we also change the average densities of the medi- um in the growth and dissolution zones. This change affects the oscillation period of the system. Thus after changing the thermal conditions we must once again carry out a "zero" determination of the oscillation period.

2. If we do not propose to make any substantial change in the mode of crystal growth during the experiment, we may con- sider another method of observing the kinetics of the process. In setting up the experiment we place weights on the lower shelf of the counterweight. After bringing the autoclave into its working state we measure the "zero" oscillation period, and, by moving weights to the upper shelf (simulating growth) plot a graph of the

reduction in the oscillation period as a function of the motion of the mass (introducing a formula correction into the graph). By returning the weights to the lower shelf (the zero reading should be restored if the crystals have not grown very much during this period), we shall be able to follow the growth of the crystals by noting the increase in the oscillation period as time passes.

An apparatus constructed on the lines indicated enables us to measure the temperature distribution in the autoclave periodically by means of a sliding thermocouple, i.e., to know the temperature difference for which crystals grow at any particular growth rate.

In order to ensure that these considerations should be completely objective, it is important to make sure that the mass transfer in the autoclave is sufficiently great, i.e., that the amount of fresh material passing into the growth chamber does not restrict the growth of the crystals. The apparatus depicted in Fig. 5 enables us to determine the mass transfer from the power consumed by the lower heater, but only provided that, by making a preliminary experiment with the same temperature distribution (without any convection), we have measured the power required by the lower heater in the empty autoclave, which under these conditions is equal to the sum of the thermal losses from the lower part of the apparatus, and should not be included when determining the mass transfer in the autoclave by reference to the power consumed [1].

Devices of this kind enable us to determine the growth rate of a particular face of the crystal as a function of ΔT for a known (but different for every value of ΔT) mass transfer, all in a single, but fairly prolonged experiment.

PVT Method of Following the
Growth of Crystals in Steady-State
Autoclaves

When growing crystals by the temperature-drop method, the charge is initially placed in the lower, "hotter" part of the liquid-filled autoclave, while the seed crystals are placed in the upper part. The convective circulation of the solution causes the gradual dissolution of the charge and the deposition of the latter in the form of single crystals on seeds placed in the upper section. The crystalline material moves toward the upper part of the vessel, while

the solution moves from the "cold" to the "hot" part of the vessel
as the crystals continue growing. Since the thermal expansion
coefficient of liquids is greater than that of crystals, this displacement of the solution should be accompanied by an increase in the
pressure, and hence the growth of the crystals may, in principle,
be followed by studying the rise in pressure in the autoclave for
a constant temperature and temperature distribution.

The pressure dependence of the increase in the volume of the
growing crystals may be described by the equation

$$V_{cr} = \frac{V_{au} \Delta P}{2\Delta T K} \, ,$$

where V_{cr} is the volume of the growing crystals, V_{au} is the volume
of the autoclave, ΔP is the change in pressure caused by the growth
of the crystals, ΔT is the temperature difference between the dissolution and crystal-growth zones, and K is the pressure increment
for the isochoric heating of the solution through 1°C at the temperature and pressure of the process.

For a pressure of 350 atm and a temperature of 350°C the
value of K is approximately equal to 10 atm. For higher temperatures and pressures it rises to 30 atm and may be measured in
every experiment as the autoclave is being brought into the crystal-growth mode.

Let us estimate the requirements which will be laid upon the
sensitivity of the manometer if we should wish to "commission"
it with the task of following crystal growth. Let us suppose that
in an autoclave of capacity 500 liters (or ml) we propose growing
crystals with a total volume of 50 liters (or ml) over a period of
20 days at a pressure of 350 atm and a temperature of 350°C. The
coefficient K will be equal to 10. If the crystals grow for a temperature difference (ΔT) of 10°C, the total rise in pressure over
the whole cycle will be

$$\Delta P = \frac{2\Delta T K V_{cr}}{V_{au}} - \frac{2\cdot10\cdot10\cdot50}{500} = 20\,\text{atm}.$$

Hence the pressure will change by 1 atm every day. In order to be
able to establish the crystal growth reliably every day, the ac-

curacy of the pressure measurement should be of the order of 0.2 atm, so that extremely accurate manometers have to be employed.

The problem is simplified by the fact that only the pressure fluctuations and not the absolute pressure have to be measured to this accuracy. For measurements of this kind we may use an ordinary manometer with a Bourdon tube free from any mechanical contact with a moving pointer, since any coupling of this kind would greatly reduce the accuracy of small readings. The small elastic displacements of the end of the Bourdon tube (which may be observed in a microscope with a magnification of 100) reliably and reproducibly indicate pressure changes to an accuracy of 0.1 atm.

Since the pressure in the autoclave depends strongly on temperature (in our own case 10 atm per 1°C), the accuracy of the temperature measurement at the moment of reading the pressure should be 0.01°C. Here also only a relative accuracy is required in order to determine the corrections to be made to the pressure readings taken at slightly differing temperatures. An ordinary thermocouple with a well-thermostated cold junction (melting point of ice obtained from distilled water) and a sensitive millivoltmeter provide sufficient accuracy in following the temperature. A resistance thermometer gives the required accuracy even more easily.

A description of an autoclave system suitable for following the growth of crystals by reference to changes in pressure was presented in [1].

The manometer for recording the small pressure changes should be set in the control desk in which the temperature of the crystal growth and dissolution chambers in the autoclave is measured at the same time. Knowing the increase in the volume of the crystals grown over a set period of time and the area of the seeds placed in the autoclave, we may easily calculate the crystal growth rate and regulate the latter by varying the temperature difference between the growth and dissolution zones.

This method of following the crystal growth rate may clearly be used most effectively when developing growth processes for particular crystals at the stage of preparation for industrial production, during the industrial assimilation of the crystal growth process, and also in installations employed for finding new techniques of producing more perfect crystals under industrial conditions.

Literature Cited

1.   A. A. Shternberg, in: Hydrothermal Synthesis of Crystals [in Russian], Izd.
     Nauka (1968), p. 203
2.   R. A. Laudise, J. Phys. Chem. Solids, Suppl. 1 (1967).

Apparatus for Precision Research in Hydrothermal Experiments

N. Yu. Ikornikova, A. N. Lobachev, A. R. Vasenin,
V. M. Egorov, and A. V. Antoshin

A great deal of work on the construction and manufacture of new apparatus has been carried out over a number of years in the Hydrothermal Synthesis Laboratory operating in conjuction with the Special Designs Office of the Institute of Crystallography.

Constant improvements to the apparatus have been called for by the desire to secure more accurate parameters in the study of various systems, and also by the need to create installations in which the growth of the crystals might be regulated at will. In order to ensure reproducibility of the results and control of the whole process, new developments are required in a number of aspects of technology. This applies in particular to material technology, to the construction of autoclaves, particularly as regards their closures, and also to the technique of thermal regulation.

In this article we shall give a description of certain new constructions of the closures and thermal regulators used in the building of our experimental apparatus. We consider such apparatus as being most suited to the synthesis and growth of crystals in aqueous solutions at high temperatures and pressures, as well as to general physicochemical investigations.

Closures

Various types of closure which have been developed in the Laboratory on the principle of compensated and uncompensated areas were described in an earlier article by Litvin and Tules [1].

In proposing new versions of closures we have based our considerations on the need not only to close the inner space of the autoclave
hermetically but also to isolate it from the steel parts of the system. For this purpose the inner space of the autoclave is lined
with titanium. This lining is produced by hot press-fitting the
titanium on to the steel, or by cold press-fitting under a high-pressure press. Initially we tried using an obturator entirely made
of titanium or titanium alloy; however, we found that this kind of
obturator gradually distorted under pressure and became unser-
viceable. We therefore made obturators of ÉI-437B steel and
protected them with a titanium layer. Obturators of this type lasted
for a long time.

In physicochemical experiments in which an exact determination of volume is required, it is best from our own point of view
to use a closure with a cylindrical type of housing and an obturator
with a copper or Teflon sealing ring. The construction of this type
of closure is based on the principle of an uncompensated area.
We used such closures in our experiments for constructing PTFC
diagrams and when studying the solubility of crystals, as well as
in standard autoclaves with a volume of about 200 cm^3 used for
synthesizing materials by transport recrystallization.

In order to grow crystals in large volumes we developed a
new type of obturator in which the sealing part consists of alternating
steel and Teflon rings (Fig. 1). The sealing principle lies in the
fact that the Teflon rings flow out into the gap between the surface
of the housing and the obturator, while the steel rings close the
Teflon and prevent it from flowing away. The lower Teflon ring

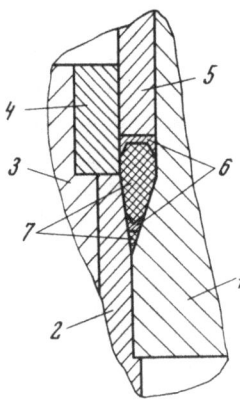

Fig. 1. Teflon sealing arrangement. 1) Obturator 1; 2) titanium lining; 3) autoclave
body; 4) ring of ÉI-437B steel; 5) pressure
ring; 6) closing ring, 7) Teflon rings.

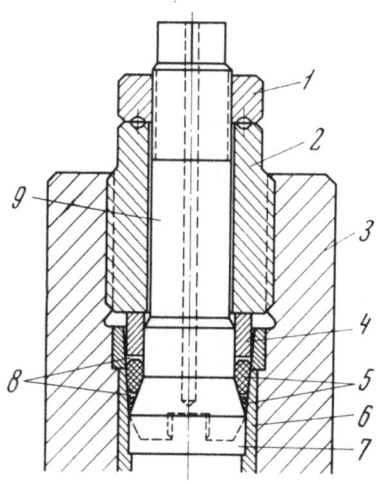

Fig. 2. Closure with Teflon seal for a 1 liter autoclave (housing diameter 50 mm). 1) Nut for opening the autoclave; 2) pressure nut; 3) autoclave body; 4) pressure ring made of ÉI-437B steel; 5) Teflon rings; 6) titanium lining; 7) titanium cap for obturator; 8) closing rings; 9) obturator.

flows out into the gap between the titanium lining of the housing and the obturator, protecting part of the obturator and the steel ring from contact with the solution. The upper Teflon ring is the principal closing layer. In this closure construction an SF-4 polyfluoroethylene seal is suitable up to 450-500°C.

For obturators with sealing Teflon and metal closing rings, nuts of different constructions are employed, depending on the autoclave housing diameter. In autoclaves with a volume of 1 liter and a cavity diameter of 5 cm sealing is effected with an ordinary nut through a steel ring (Fig. 2). In autoclaves of diameter 70 mm (volume 4 liters) the closing rings are pressed in by means of a bushing and screws (Fig. 3).

In autoclaves with an internal diameter of 100-200 mm closures without sealing rings are employed. The hermetic sealing of these closures is ensured by carefully polishing the surface of the obturator and the fitting point.

Apparatus for Electronic Thermal Regulation

Our research called for the development and manufacture of thermal regulators with a fair degree of reliability over prolonged periods of use and of small dimensions. We developed two new circuits based on proportional regulation (Fig. 4). These circuits were designed to maintain the sample temperature constant at

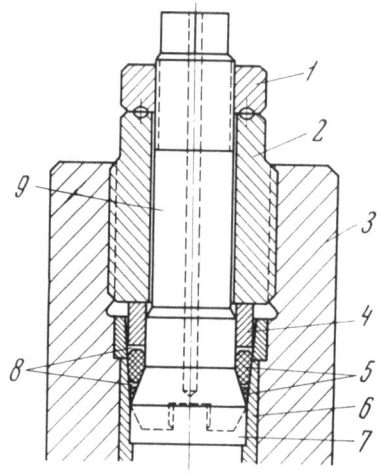

Fig. 3. Closure with Teflon seal for
a 4 liter autoclave (housing diameter
70 mm). 1) Obturator; 2) disc for re-
moving obturator; 3) pressure screw;
4) autoclave body; 5) pressure bushing;
6) Teflon rings; 7) titanium cap; 8)
closing rings; 9) bearing nut.

points between 100 and 600°C; they consist of a self-balancing bridge
and an executive mechanism. A vital sensitive element of these
circuits is a platinum resistance thermometer serving as sensor.
Various types of sensor are used: an open type up to 400°C and a
closed type (in a quartz sheath) up to 600°C. Later we used resistance elements of the LT-17 9855 type of laboratory manufacture
as sensors, these being distinguished by excellent insulation.

In the first version of the circuit (Fig. 4a) a platinum resistance
thermometer forms one arm of the measuring bridge. Another
arm of the bridge consists of a series resistance box and slide
wire for the coarse and fine balancing of the bridge. The third arm
includes a slide wire for the automatic compensation of disbalance.
The bridge is supplied with a d.c. voltage of 8 v from a rectifier.
The nul device is a UÉU-109 amplifier which governs a reversing
RD-09 motor. The latter is kinematically coupled to the axle of
the slide wire used for automatically compensating the disbalance
of the bridge, and through a reducing gear to the executive mechanism, which in the present case constitutes a linear autotransformer ('Latr").

In the second version (Fig. 4b) a magnetic amplifier of the
UM-1P type is used as executive mechanism.

Small-scale units for one, two, and four heaters were con-

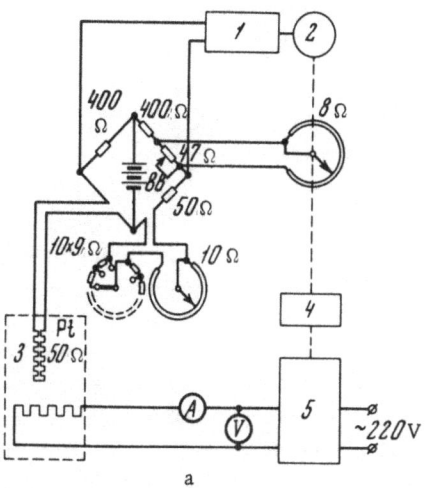

a

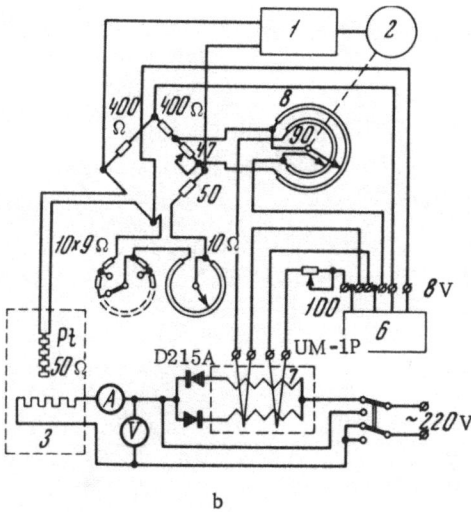

b

Fig. 4. Circuits of small-scale electronic regulators. 1) UÉU-109; 2) RD-09; 3) furnace; 4) reducing gear; 5) autotransformer; 6) supply source;
7) magnetic amplifier.

structed and manufactured in the Laboratory on the basis of these circuits.

Installations for Physico-chemical Investigations and Growing Crystals

1. Installations for Plotting PTFC Diagrams by the Isochore Method. The devices in question comprise an autoclave with a manometer, heaters, and a thermal regulator. The autoclave, 160 cm³ in volume, with a cylindrical closure, is made of Kh18N9T stainless steel; it is connected to the manometer through a capillary and an intermediate vessel. The pressure is measured with standard manometers (1000, 1500, and 2500 kg/cm²) calibrated in the Institute of Standards amd Measuring Instruments. The autoclave is heated with three independent heaters. The principal heater is wound on the cylinder (brass or copper) in which the autoclave is placed. Around the cylinder and heaters is a thermally-insulated zone. The two other heaters are used for compensating the temperature drop along the autoclave. Along the cylindrical walls of the furnace as well as the top and bottom are water-cooling zones. The temperature is measured at three points on the autoclave body, the thermocouples being inserted into channels drilled in the body so as to bring the hot junctions of the thermocouple to within 3 mm of the solution zone. The temperature in the body of the autoclave is made uniform by means of the three heaters, varying by no more than 1° from point to point.

In the second version of the apparatus for plotting PT curves (Fig. 5) the ÉPR-09-RD regulator was replaced by an electronic regulator having a resistance thermometer as sensor, installed in the thick-walled brass housing of the furnace. The temperature fluctuations in this system never exceed ± 0.1° [2].

2. Installation for Studying Solubility. Installations for studying solubility are depicted in Figs. 6 and 7. One of these is constructed for working with a single autoclave and the other for working with five. For studying solubility small autoclaves (25-50 cm³) are used, without any manometer. The furnaces accommodating the autoclaves are characterized by isothermal heating. In the furnace designed for a single autoclave, the heater is wound on a brass cylinder with a cover plate, and the autoclave is placed in this. The gap between the walls of the autoclave and

the cylinder is no greater than 0.3 mm. The thickness of the cylinder walls is 12 mm, the thickness of the top 50 mm. The thin-
walled brass sheath equalizes the temperature over the autoclave
surface, and no compensating heaters have to be applied. A channel is drilled in the wall of the brass cylinder for a resistance
thermometer (sensor), which is connected to an electronic regulator. In the brass body of the heater the temperature is held to an
accuracy of ± 0.1°.

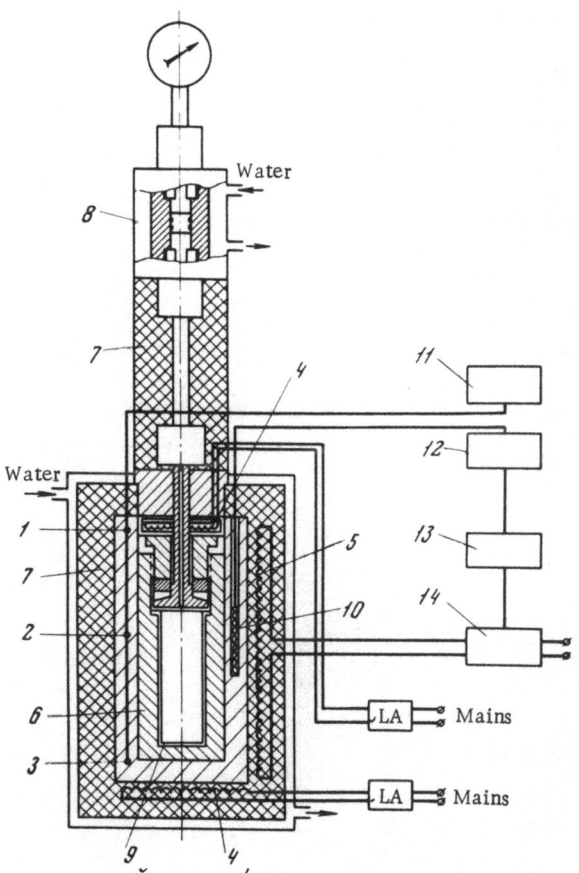

Fig. 5. Apparatus for plotting PTFC diagrams. 1), 2), 3)
Thermocouples; 4) compensation heaters; 5) principal heater; 6) autoclave body; 7) thermal insulation; 8) barrel with
piston; 9) titanium lining; 10) platinum resistance thermometer (sensor); 11) potentiometer; 12) measuring bridge;
13) proportional regulator; 14) autotransformer. LA = linear
autotransformer.

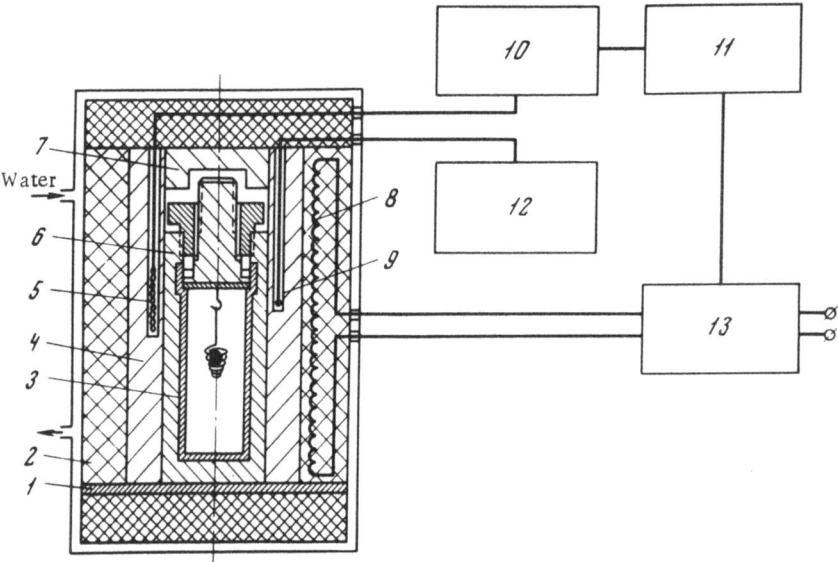

Fig. 6. Apparatus for studying solubility in one autoclave. 1) Asbestos cement slab; 2) thermal insulation; 3) autoclave lining; 4) brass cylinder; 5) platinum sensor; 6) autoclave body; 7) brass top; 8) body heater; 9) thermocouple; 10) measuring bridge; 11) proportional regulator; 12) potentiometer; 13) autotransformer.

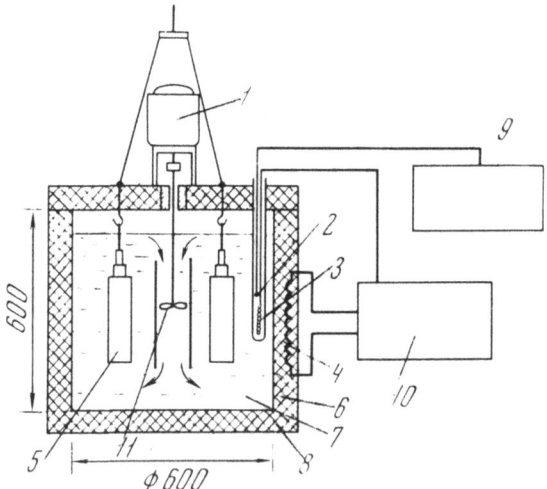

Fig. 7. Apparatus for studying solubility (liquid thermostat). 1) Motor; 2) thermocouple; 3) sensor; 4) heater; 5) autoclave; 6) body of furnace; 7) salt solution; 8) thermal insulation; 9) potentiometer; 10) electronic regulator; 11) stirrer.

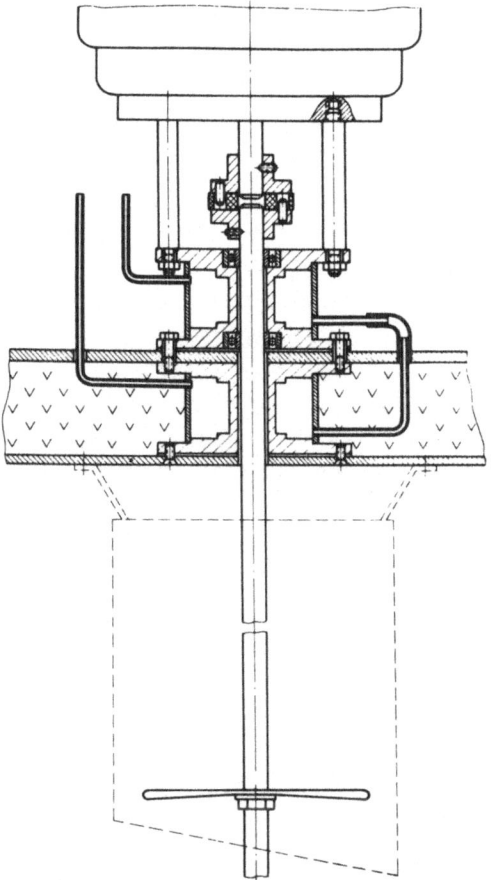

Fig. 8. Motor and stirrer of the liquid thermostat.

The furnace for determining solubility simultaneously in five autoclaves has the form of a cylinder made of stainless steel sheet and filled with a nitrite–nitrate mixture (40% $NaNO_2$, 7% $NaNO_3$, 53% KNO_3). The mixture has a melting point of 140°. In the center of the cylinder is the rod of a paddle stirrer; around this are the autoclaves and the sheath for the sensor. The autoclaves are placed vertically and held stationary. The rod of the stirrer and the motor driving it are fixed in the top of the furnace. During the experiment the autoclaves and sensors are let down into the molten mixture, which is agitated vigorously. In other words, the construction constitutes a high-temperature liquid thermostat with a working

temperature range of 150–550°C (for the composition indicated).
By using this thermostat and carrying out a single experiment we
may obtain enough points to construct the solubility isotherms.

For agitating the solution (in the single-phase state with the
autoclave set vertically) the magnetic stirrer earlier described in
[3] may be employed. The autoclave with the magnetic stirrer
(Fig. 9) may be used in the proposed version of the group furnace.

In the second version of the system the autoclaves are placed
horizontally in a hinged frame attached to a shaft with a cam. This
construction is intended to provide a periodic change in the inclination of the autoclaves to the horizontal plane. A platinum
resistance thermometer (or a set of thermocouples) in a ceramic
shield is placed in a stainless steel tube. Through an opening in
the top this is let down into the salt melt and connected to an elec-

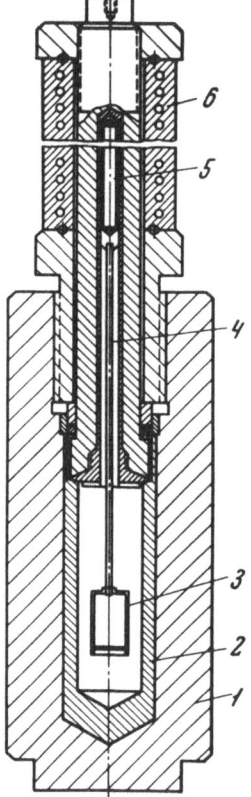

Fig. 9. Autoclave with magnetic stirrer.
1) Autoclave body; 2) lining; 3) frame
for crystals; 4) titanium rod; 5) mild steel
rod; 6) solenoid with thermal insulation.

tronic regulator. The temperature fluctuations in the thermostat
amount to no more than ± 0.1° throughout the experiment. The
furnaces of the solubility-measuring systems are left switched on
throughout the whole cycle of experiments, which is conducted at
constant temperature.

3. Installation for Synthesizing Crystals
(Group Furnace). This installation enables a whole group of
autoclaves to be heated at the same time and consists of a furnace
and an electronic regulator. In the furnace the autoclaves are placed
on a cast iron slab with a Nichrome strip heater in ceramic placed
under it. This type of heating element is very simple and provides
reasonably uniform temperature conditions in all the autoclaves.
The heating temperature is regulated by means of a chromel–
alumel thermocouple sensor in a porcelain tube placed in a socket
drilled in the center of the slab. The porcelain tube is set vertically in the center of the slab and passes through the top. The ther-

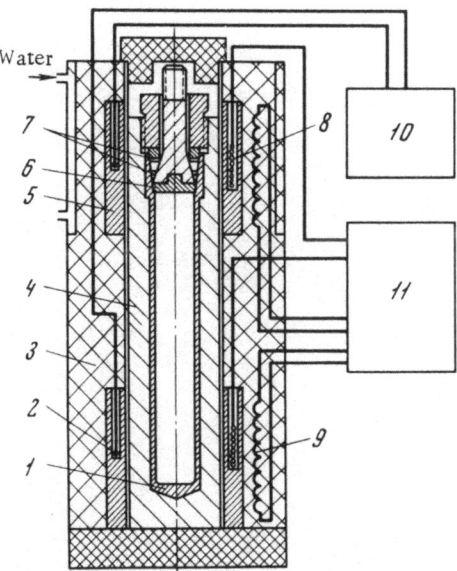

Fig. 10. Arrangement of autoclave with heaters
and a thermal regulator. 1) Titanium lining; 2)
thermocouples; 3) thermal insulation; 4) autoclave
body; 5) brass or copper cylinder; 6) titanium disc;
7) Teflon seal; 8) resistance thermometer (sensor);
9) heater; 10) recorder; 11) electronic regulator.

mocouple is connected to a regulating and recording electronic
potentiometer, the signal from which passes to a proportional
regulator based on magnetic amplifiers. The regulator circuit
enables the temperature in the cast iron slab to be kept constant to
within ± 0.5°. An upper heater in the top of the furnace is provided
in order to vary the temperature drop.

4. Installations for Growing Single Crystals.
These installations constitute individual autoclaves with zonal or
continuous heating and electronic thermal regulation units. Each
autoclave is equipped with platinum or chromel−alumel sensors,.
inner measuring thermocouples, cooling jackets, manometers,
and high-pressure valves. Figure 10 illustrates a 1.2-liter autoclave with two zone heaters. The heaters are wound on brass
frames of 10 mm thick with pockets for the sensors. Both sensors
are connected to a unit with a two-channel electronic regulator
(Fig. 11). The autoclave is machined from 45KhNMFA steel, lined
with ST-1 titanium, and has an obturator with a protective titanium

Fig. 11. General view of a two-channel
electronic regulator (dimensiones 700 ×
420 × 320 mm).

disc. The construction of the autoclave envisages a pressure of
1800-2000 kg/cm^2 at 500°C.

An autoclave made of Kh18N10T steel lined with ST-1 titanium
has also been made; this has a volume of 4 liters and is designed
for a pressure of 400 kg/cm^2 at temperatures up to 300°C.

Figure 12 illustrates an installation for studying the growth of
crystals in solution at high temperatures and pressures under temperature-drop conditions. The autoclave is made of Kh18N10T
steel and has a volume of 9 liters, the diameter of the inner space
being 90 mm. The solution, seed, and charge are placed in a glass
inner vessel 5 cm in diameter supported on a Teflon stopper. Inside the vessel the growth and dissolution zones are separated by
a Teflon barrier. The inner vessel is set in a brass block so that
the end of the block lies at the level of the Teflon barrier. A thermocouple pocket made of Kh18N10T steel is placed in a channel
drilled in the block.

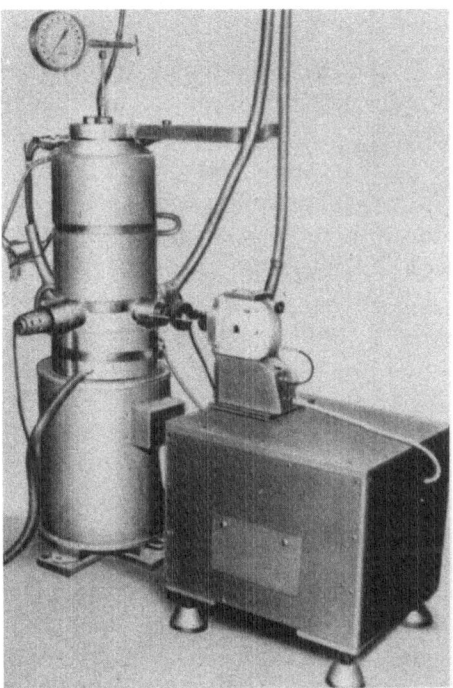

Fig. 12. General view of an autoclave with
windows and a motion-picture attachment.

The inside of the autoclave surrounding the upper part of the inner vessel is filled with an inert gas to balance the pressure developing in the inner vessel over the solution at high temperatures.

Three apertures are drilled in the body of the autoclave and tubes are sealed into these. In the tubes are transparent windows and seals, the windows consisting of two cylinders, one made of fused quartz and the other of leucosapphire, connected together by the optical contact of their bases. The surface of the leucosapphire is turned toward the inside of the autoclave.

In the body of the autoclave three heaters are mounted on brass frames. Sensors are provided by platinum resistance thermometers connected to a three-channel electronic regulator. Temperature fluctuations in the heaters are no greater than $\pm 0.1°$, and inside the autoclave there are hardly any fluctuations at all. The temperature measured with chromel–alumel thermocouples along the walls of the glass vessel and in the brass (inside the autoclave) is recorded with an ÉPP-09-RD potentiometer.

The crystal is photographed during growth by means of a motion-picture camera, with interchangeable tubes for taking photographs at different magnifications. At one end of each tube is a connecting piece fixed in the entrance window of the camera. At the other end in the objective, which is set in motion by means of a rack and pinion. An automatic device for periodically photographing the growth of the crystal enables frame-by-frame pictures to be taken at between one frame in two seconds and one frame per hour, intermediate frequencies being related to each other by a factor of two. At the same time the device enables illumination to be provided with a lead of 0.5 sec or over, starting from a frame frequency of one per minute or slower. The camera is started by hand.

Conclusion

The new apparatus described in this article is being used in a number of experimental investigations. The development of new electronic circuits for thermal regulation enables us to keep temperatures constant in furnaces to an accuracy of $\pm 0.1°$, or in some cases $\pm 0.5°$. The thermal-regulation units based on these circuits have proved convenient under laboratory conditions and are completely reliable. The use of the thermal regulators here considered in installations for growing crystals has ensured excellent reproducibility of the results. In the field of physicochemical research, the better degree of thermal regulation thus achieved and the use of new lining materials have minimized the scatter of experimental points on the monovariant curves associated with the determination of various parameters. This has revealed, for example, the different ways in which chloride solutions behave when the parameters P, T, F, and C are varied, and has also enabled us to differentiate the solubility curves of crystals in different solutions, both as regards their general character and also as regards their extremal values.

In this way we have been able to establish the individual characteristics of the solubility diagrams of crystals in various alkali chloride solutions.

The group furnace (liquid thermostat) for studying solubility differs considerably from those used earlier, both as regards the construction of the furnace and also as regards the accuracy of parameter determination. A furnace in which the autoclaves are heated in a well-stirred salt melt is more isothermal than one in which the autoclaves are heated in an air space.

Using an autoclave with windows, we have been able for the first time to study the growth of a crystal in aqueous solutions at temperatures up to 200°C and pressures up to 300 kg/cm^2.

Literature Cited

1. B. N. Litvin and Zh. A. Tules, in: Hydrothermal Synthesis of Crystals [in Russian], Izd. Nauka (1968), p. 193.
2. N. Yu. Ikornikova and B. M. Egorov, in: Hydrothermal Synthesis of Crystals [in Russian], Izd. Nauka (1968), p. 58.
3. D. S. Tsiklis, Technique of Physicochemical Investigations at High Pressures [in Russian], Izd. Khimiya (1965), pp. 236, 252.